BIASED TECHNICAL CHANGE AND ECONOMIC CONSERVATION LAWS

Research Monographs in Japan-U.S. Business & Economics

series editors

Ryuzo Sato
Rama V. Ramachandran
Stern School of Business
New York University
Kazuo Mino
Kobe University
Japan

Other books published in the series:

Sato and Ramachandran
Conservation Laws and Symmetry: Applications to Economics and Finance

Sato, Ramachandran, and Hori
Organization, Performance, and Equity: Perspectives on the Japanese Economy

Sato, Grivoyannis, Byrne, and Lian
Health Care Systems in Japan and the United States: A Simulation Study and Policy Analysis

Sato and Ramachandran
Symmetry and Economic Invariance: An Introduction

Sato, Ramachandran, and Mino
Global Competition and Integration

Negishi
Developments of International Trade Theory

Ihori and Sato
Government Deficit and Fiscal Reform in Japan

Hamori
An Empirical Investigation of Stock Markets

BIASED TECHNICAL CHANGE AND ECONOMIC CONSERVATION LAWS

by

Ryuzo Sato
C.V. Starr Professor of Economics
and Director of the Center for Japan-U.S.
Business and Economic Studies
Leonard Stern School of Business
New York University

Springer

Library of Congress Cataloging-in-Publication Data

Sato, Ryuzo
Biased technical change and economic conservation laws
p.cm. (Research monographs in Japan-U.S. business & economics; 9)
Includes bibliographical references, and index.

ISBN 0-387-26055-2 e-ISBN 0-387-26376-4
ISBN 978-0387-26055-6

Printed on acid-free paper.

Printed in the United States of America.

9 8 7 6 5 4 3 2 1 SPIN 11424000

springeronline.com

To my grand children:
Keimi, Yūki, Ryuka and Yūna

Acknowledgements

The publishers wish to thank the following who have kindly given permission for the use of copyright material.

1. Blackwell Publishers for article: "Optimal Growth with Endogenous Technical Progress: Hicksian Bias in a Macro Model," (Chapter 4 of this book), with Rama Ramachandran, *The Japanese Economic Review*, 51 (2), 2000, 193–206.
2. Cambridge University Press for articles: "Model of Optimal Economic Growth with Endogenous Bias," (Chapter 3 of this book), with Rama Ramachandran and Chengping Lian, *Macroeconomic Dynamics*, 3, 1999, 293–310;
   "Optimal Economic Growth: Test of Income/Wealth Conservation Laws," (Chapter 8 of this book), *Macroeconomic Dynamics*, 6, 2002, 548–572.
3. Edward Elgar for excerpt: "Dynamic Symmetries and Economic Conservation Laws," (Appendix of this book), *Theory of Technical Change and Economic Invariance: Application of Lie Groups*, Chapter 7, Sections I, II & III, 233–259.
4. Elsevier for articles: "International Competition and Asymmetric Technology Game," (Chapter 7 of this book), *Japan and the World Economy*, 13, 2001, 217–233;
   "Hartwick's Rule and Economic Conservation Laws," (Chapter 11 of this book), with Youngduk Kim, *Journal of Economic Dynamics and Control*, 26, 2002, 437–449;
   "Economic Conservation Laws as Indices of Corporate Performance," (Chapter 9 of this book), *Japan and the World Economy*, 16, 2004, 247–267;
   "Conservation Laws for Microeconomists! Coment" by Thomas Mitchell (Appendix to Chapter 9 of this book), *Japan and the World Economy*, 16, 2004, 269–276;
   "Evaluating Corporate Performance: Empirical Tests of a Conservation Law," (Chapter 10), with Mariko Fujii, *Japan and the World Economy*, forthcoming 2005.
5. Keio Economic Studies for article: "A Note on Modelling Endogenous Growth," (Chapter 6 of this book), *Keio Economic Studies*, 33 (2), 1996, 93–101.
6. Kluwer Academic Publishers for excerpt: "Estimation of Biased Technical Progress," (Chapter 5 of this book), with Rama Ramachandran and Youngduk Kim, in *Global Competition and Integration*, ed. by Ryuzo Sato, Rama Ramachandran and Kazuo Mino, 1999, 127–170.

Preface

Of the two topics mentioned in the title of this book, *Biased Technical Change and Economic Conservation Laws*, the former deals with an issue that has been frequently discussed in the existing literature. The latter, on the other hand, is still relatively unfamiliar to many economists. They both have one thing in common, however: they have basically never been thoroughly analyzed before.

Although terms such as "labor saving" and "capital saving" fall under the category of biased technical change, the first of these topics, no analysis has ever been made of its basic statistical estimation methodology or its structure relative to production functions. In particular, no model exists in which biased technical change gives rise endogenously to technical progress. A special feature of this book is its thorough investigation and analysis of these issues, which go far beyond existing studies in this area.

The concept of economic conservation laws dates back to Ramsey's classic study of 1928, but until recently no one has ever dealt with its present-day significance or its application to economics attendant upon its subsequent development in the other social sciences. This book primarily makes use of Lie groups to shed new light on the analysis of economic conservation laws. Economic conservation laws are not simply abstract concepts; this book shows that they are tools of empirical analysis that can be applied to such topics as analyses of macro performance and corporate efficiency.

In short, the principal aim of this book is twofold: to reveal the new economic significance of the old concept of biased technical change and the current application of the new concept of economic conservation laws.

Many of the chapters here are revised versions of papers that were originally published in academic journals, subsequent to the publication of Volume One and Volume Two of *The Selected Essays* of Ryuzo Sato (Edward Elgar, 1996 & 1999). This volume is, in a sense, Volume Three of my *Selected Essays*. I owe a great debt of gratitude to my co-authors of several articles in this volume, Rama Ramachandran, Chanpin Lian, Youngduk Kim and Mariko Fuji. I would also like to thank Blackwell, Cambridge University Press, Edward Elgar, Elsevier, Kluwer Academics, and *Keio Economic Studies* for permitting me to include these previously published works in this book.

New York Ryuzo Sato

Contents

Chapter 1

AN OVERVIEW

1. Introduction

1. This book deals with the two topics cited in the title: biased technical change and economic conservation laws. At first glance, the two might appear to have no connection with one another, yet when viewed from the perspective of the optimal control behavior of the primary agents in an economic analysis, both topics have much in common and are profoundly related as phenomena that are the result of optimized behavior.

The term "biased technical change," needless to say, refers to a situation in which the factors of production (in this case, capital and labor) each achieve technical change at different rates. As a result, the efficiency of capital and the efficiency of labor exhibit different growth rates. Consequently, even though the ratio of capital to labor remains constant, the shares of each are affected by technical change. This is the essence of biased technical change.

2. The accepted postulate among economists has been that, as a special case of biased technical change, only Harrod neutral technical change is valid as a condition for balanced growth in a mature economy. In other words, for long-run macroeconomic stability, technical change must of necessity be Harrod neutral technical change. To be more precise, this means that the only hypothesis conducive to long-run balanced growth is one in which capital-augmenting technical change does not exist.

What economists have learned from serious empirical analysis over the past few decades, however, is that capital-augmenting technical change is not zero. On the contrary, it varies significantly in an economic upswing or downturn, increasing greatly during the former and decreasing during the latter. The trend is not zero, however. In short, technical change during an upswing is not completely canceled out by technical change during a downturn. Consequently, Harrod neutral technical change, which assumes a zero rate of capital-augmenting technical change not just over the short run but even in the long run, is an unrealistic hypothesis. It is merely the simplest mathematical hypothesis to ensure long-run macroeconomic stability.

3. It makes no sense to regard advances in information technology, computer functions, for example, which have been growing by leaps and bounds, as all the result of a rise in Harrod neutral labor efficiency. In the ten years that the author has used a computer, the technical change of his labor has remained virtually the same. Nevertheless, it is a fact that every time I buy a new computer, the efficiency of my work rises rapidly.

This fact is readily analyzable if one simply accepts that the computer's capital efficiency has risen. In order to explain why the author is able to make more expeditious use of the computer while his labor efficiency remains unchanged, an indirect explanation-the assumption that new computers are improving because the productivity of those workers involved in manufacturing them has gone up-is far-fetched. This would mean believing indirectly in a Ricardo-Marxian labor theory of value in which all values derive from labor.

4. If labor and capital are regarded as independent production factors, a company/agent might focus its energy on increasing the efficiency of labor over that of capital, depending on the economic situation at the time. Or, faced with too little capital, it might invest more resources into increasing the efficiency of capital rather than labor.

At the time of the oil crises of the 1970s, companies confronted by soaring oil prices focused on technical change as a way to economize on capital, i.e. oil. Indeed, technological development was precisely what saved Japan's resource-poor economy. In Japan, incidentally, this is called "sho-ene" or energy-saving technology development.

5. The differences between this book and other works are (1) its main theme is an analysis of precisely this sort of technical change in which the efficiencies of capital and labor rise at different rates, and (2) it shows that long-run economic stability is possible under a biased technical change other than Harrod neutral. The basic concept here is that, of the different types of technical change, capital-augmenting technical change occurs endogenously. To be more precise, the theory that an economy achieves stable balanced growth only under the existing Harrod neutral technical change is limited to a situation in which technical change of labor and of capital is exogenous; in short, the theory applies only to those situations in which technical change is bestowed upon a nation or one of its companies by another nation or another company or by Heaven without that nation or its company using any resources of its own. But technical change ought to be regarded as attainable only when investment is made in resources. This is the theory of endogenous technical change put forward in this book.

6. The essence of the theory of endogenous technical change is the question of optimization: i.e. how much of the resources that a nation or a company possesses should it invest in technical change? Specifically, this book attempts to analyze this problem using the Pontryagin-Hesteness-Bellman calculus of variation or an optimal control method.

7. The theory of economic conservation laws that makes up the second theme may be less familiar to many economists than the first concept of biased technical change. But in the fields of physics, applied mathematics and other modern sciences, it is becoming a tool for studying the question of optimization in greater depth.

The above mentioned calculus of variation or optimal control theory is a suitable tool to explain natural or economic phenomena; behind variables and systems that are mathematically observable are phenomena that cannot at first glance be observed. To take a well-known example in physics, there is a rule (conservation law) that, while the movement of a billiard ball is observable, the sum of the kinetic energy and the potential energy behind it, which are impossible to observe, is constant; this rule is constancy of the Hamiltonian function.

8. In endogenous occurrences of biased technical change or in the maximization of a company's long-run profits, we are able to observe fluctuations over time in technical change rates or in profit rates. Likewise, in models dealing with a country's optimum capital savings or its optimum technical change rate, we can observe variables in savings and investment, consumption and GDP, just as we can observe the movement of a billiard ball, but, of course, the more essential but invisible and hidden law behind these optimized phenomena cannot be observed. Yet, it is this law, which the eye cannot see, that embodies the essence of the endogeneity question.

Several hidden conservation laws may exist in a single optimized system. The methodology for making a close and accurate analysis of this sort of problem is the Emy Noether theory of invariance using Lie groups. To sum up, the two main themes are bound together by the common terms endogeneity and conservation laws.

9. A Short History of Biased Technical Change. The work that formally introduced the concept of technical change to modern economics was The Theory of Wages (1932) by Sir John Hicks. Hicks analyzed the impact of inventions and technical change on the shares of labor and capital in Britain's post-industrial-revolution economy and society. Even in situations where the ratio of capital to labor was constant and unchanging, if technical change occurred, it might have an effect on the shares of capital and labor or their distributive shares.

Even though the ratio of capital to labor remained unchanged, a new invention might make the share of capital increase and the share of labor decline. Or the opposite might occur. As the simplest case, Hicks conceived of a situation in which a new invention or technical change had absolutely no impact whatsoever on the shares of labor and capital. To be more precise, one in which, under a constant capital-to-labor ratio, a new invention or technical change would have no effect on the income distributions of labor or capital, or the effect would be neutral. This subsequently came to be called Hicks neutral technical change.

10. The 1930s when Hicks advanced his theory of technical change was the era of the Great Depression, which had started in the United States and spread throughout the globe. The world's economists paid no attention to Hicks' contribution to the analysis of questions related to growth and technical change. Keynes' General Theory came out in 1936, a few years after the publication of Hicks' book, at the very time when an academic theory of how to get through the depression was in high demand. Needless to say, Keynes' work attracted the attention not only of economists but of politicians and policy-makers.

After World War II at the beginning of the 1950s, economists primarily in the United States began to think that, by skillfully combining Keynes' prescription and market principles, it would be possible for the world economy to escape recession and achieve stable growth. This was the theory of a mixed economy set forth by Paul Samuelson.

11. By the mid 1950s the belief had become prevalent that the world economy, and especially the American economy, would never experience another Great Depression like the one in the 1930s. On the contrary, given the prosperity and growth of the US economy resulting from postwar technological advances, there was a growing view that, in addition to the growth of labor and capital, reconsideration needed to be given to the significance of the contribution of technical change. For the first time, the theory of technical change that Hicks had analyzed in the 1930s began to attract economists' interest. Representative of this trend was Professor Robert Solow of MIT.

12. Solow analyzed growth trends in the non-agricultural private sector using Kendrick data for the US economy between 1909 and 1949 (Review of Economics & Statistics, 1957). Using a neoclassical production function of constant returns, he hypothesized that the growth rate for income in the non-farm private sector would also be affected by the growth rate for technical change over and above the growth rates for capital and labor. The technical change that Solow used here was the very same neutral type that Hicks had analyzed in the 1930s.

Solow ascertained that part of the income growth rate which is not dependent on the growth rates of labor and capital occurs as a result of technical change. Moreover, he succeeded in estimating the growth rate of technical change simply by subtracting the sum of the growth rates of labor and capital weighted by their distributive shares from the growth rate of income. According to Solow's estimates, approximately one-third of the income growth rate is dependent on technical change. Thus, Hicks' technical change of the 1930s was revived by Solow in the 1950s.

13. Solow's achievement had an enormous impact on young economists at the time. In particular, Solow's method opened the way to estimate from the same data both the production function, which is hard to estimate, and the rate of technical change. The paper by Professor John Kendrick and the author (American Economic Review, 1963) brought this work to completion.

Using the same Kendrick data as Solow had, it was possible to estimate the technical change rate and the production function simultaneously by estimating the elasticity of factor substitution, a concept that Professor Hicks had developed in the 1930s. It was estimated that the technical change rate in the period 1909–1920 was an average of 2.1 percent, and the elasticity of factor substitution was 0.6. Thus, it was proposed that the most suitable theory for explaining the growth of the American economy in that period is a production function with an elasticity of factor substitution of 0.6 (a CES production function).

Another important point made in this article was that the estimated value of the elasticity was smaller than the Cobb-Douglas production function (=1) emphasized in the Solow estimate, i.e. a production function with an elasticity of factor subsitution of 0.6, more approximately explained the growth of the American economy.

14. There were doubts at the time whether the Hicks neutral concept might be too unrealistic to explain technical change. Economists began to perceive a need to reconsider once again the concept of neutral technical change and to analyze how technical change is related to a neoclassical production function.

In addition to the Hicks' concept of neutral technical change, in the late 1940s Harrod had studied technical change which would enable a mature economy to achieve a long-run, stable, balanced growth path, one in which a constant income-capital ratio is maintained. In other words, technical change in which the shares of labor and capital are unchanged when the ratio of income to capital is constant. This came to be called Harrod neutral technical change.

Analysis was also made of a type of technical growth which is the antithesis or mirror image of this and which applied to the economies of developing countries, i.e. one in which there is absolutely no impact on the shares of capital and labor both before and after technical change when the ratio of income to labor is constant. This is Solow-Ranis-Fei neutral technical change.

15. In the empirical field of technical change, on the other hand, a multiplicative type of technical change, in which the efficiencies of capital and labor are different, is applicable to empirical data. This is the so-called factor augmenting technical change.

Although factor augmenting technical change was known to include Hicks neutral, Harrod neutral as well as Solow-Ranis-Fei neutral technical change, the theoretical grounds under which it maintained neutrality were completely unknown. To put it another way, it was not known under what conditions the distributive shares of capital and labor would remain constant, as in the three other types of neutrality.

The First answer to this question was the article that the author and Professor Beckmann published in the Review of Economic Studies (1968). The note that Professor Rose published a year later in the Economic Journal showed the same results as the Sato-Beckmann paper. This is the theory that was later called Sato-Beckmann-Rose technical change. It is also an explanation of the above mentioned factor augmenting technical change. Hicks, Harrod and Solow-Ranis-Fei neutral are all defined as special cases of Sato-Beckmann-Rose neutral. Hicks neutral is the case in which the efficiencies of capital and labor remain the same; Harrod neutral is one in which only the efficiency of labor goes up; and Solow-Ranis-Fei neutral is one in which only the efficiency of capital goes up.

16. What is the neutrality principle that justifies Sato-Beckmann-Rose? It is this: technical change takes the form of Sato-Beckmann-Rose factor augmenting technical change if it is neutral in the sense that it has no effect on the distributive shares of labor and capital, as long as the elasticity of factor substitution remains constant. Technical change of this type is most frequently used in empirical studies; its suitability for theoretical use is explained by the fact that it is derived from the neutrality principle of technical change.

Production functions that include Sato-Beckmann-Rose factor augmenting technical change, as well as estimates of technical change and an empirical analysis had first been carried out by the author in an article in the International Economic Review of 1963. The main point of this paper was to demonstrate that, by measuring the elasticity of factor substitution on using Kendrick data and then using that value, it was possible to measure the technical growth rates of both capital and labor in two equations. In contrast to the paper by Kendrick and Sato, it explained that there is a large gap between the rate of technical change for labor and the rate of technical change for capital; hence, the Hicks' technical change hypothesis did not apply to the US economy.

It also explained that the rate of technical change for labor is higher than that of capital, and that the latter is not zero as had been assumed by Harrod neutral technical change. It might be noted in passing that the rate of technical change for labor was 2.7 percent and the rate of technical change for capital was 0.7 percent (see Sato, International Economic Review 1963).

What is noteworthy here is that, except in cases where neutrality in one model is combined with a special production function (like the Cobb-Douglas type), it does not mean neutrality in the other models. In other words, if the model adopts Hicks neutral, then by definition, technical change of the Harrod, Solow-Ranis-Fei and Sato-Beckmann-Rose types are all not neutral; in short, they are biased technical change. The technical dealt with in this book will, for the most part, be of the Sato-Beckmann-Rose type.

17. An analysis of Sato-Beckmann-Rose technical change using Lie group transformations.

Sato-Beckmann-Rose neutral factor augmenting type of technical change under a neo-classical production function of constant returns to scale is expressed as follows:

$$Y(t) = F[A(t)\ K(t),\ B(t)\ L(t)]$$
$$\bar{Y}(t) = F[\bar{K}](t),\ \bar{L}(t)]$$

Capital and labor, which we here express in terms of the efficiency of technical change, are

$$\bar{K}(t) = A(t)\ K(t)\bar{L}(t) = B(t)\ L(t)$$

This can be seen as the transformation of $K(t) \to \bar{K}(t)$

$$L(t) \to \bar{L}(t)$$

If $A(t) = A_0e^{\alpha t}$, where $\alpha \geq 0$ and $B(t) = B_0e^{\beta t}$, where $\beta \geq 0$, the transformation of K(t) and L(t) is a magnificent type of a Lie group (Sato [1981, 1999]). I will not go more deeply into this question here.

18. A short history of economic conservation Laws. Ramsey's article in the Economic Journal of 1928 was the first in the long history of economics to introduce a dynamic method, i.e. a calculus of variation. Ramsey examined the question of how much a country would need to save and invest in order to maximize welfare, which he measured in terms of a mature economy's long-run rate of consumption. This was the first attempt at an optimum savings theory, which neo-classical growth theories have often dealt with from the 1960's on.

Ramsey intentionally did not introduce the concept of the discount rate. This marks a clear distinction from the optimum savings theories of today. Consequently, he was able to use a concept similar to the one mentioned earlier, the Hamilton energy conservation law used in physics. In the law of energy conservation, at each point of a movement in time kinetic energy + potential energy = constant; similar to this, in Ramsey's case, at each point in time the sum of net welfare and the value of saving is always constant, i.e. net welfare + value of saving = constant. Moreover, this value is the highest that nation's economic system can achieve. Ramsey called this value "bliss." Total energy in the energy conservation law and Ramsey's Bliss are the values of a maximized Hamilton function in a dynamic system. In short, Ramsey knew the conservation law that, assuming a discount rate of zero, a maximized Hamilton discount rate at each point in time is always constant. Although the concept itself was not used, for all practical purposes, this was the first use of a conservation law in economics.

19. Economists did not revisit Ramsey's conservation law until the 1970s. Samuelson's paper in 1970 was the first in that era to introduce an economic conservation law into modern theoretical economics. Like Ramsey, Samuelson analyzed a von Neumann optimum problem using the analogy of the total energy law. The conclusion derived from this was the discovery that an economic conservation law is at work: aggregate capital-output ratio = constant. This was demonstrated by Sato [1981, 1999] with a detailed derivation.

Although, unlike the present book, Samuelson's article did not make use of the Noether theorem using Lie groups as his principle methodology, it was an outstanding accomplishment for both its accuracy of intuition and its economic significance.

20. Ramsey's and Samuelson's achievements were not generally known at the time the author received grants from the National Science Foundation and the Guggenheim Foundation to engage in a study of Lie groups and economic conservation laws. That was in the mid 1970s.

The author's comprehensive survey of the field, The Theory of Technical Change and Economic Invariance: Application of Lie Groups (Academic Press, 1981; revised edition, Elgar, 1999), not only made a representation of technical change using Lie groups, it conducted a thorough, full-scale analysis of economic conservation laws using the Noether theorem. It showed that Ramsey's Bliss conservation law and Samuelson's capital-output ratio conservation law are derived as special cases of the Noether theorem. (The Noether theorem, simply put, is that if a dynamic system including integral calculus is constant under Lie transformation groups that have r-parameters, r conservation laws exist in it).

The Ramsey model and other neo-classical long-run growth models were analyzed in detail in Chapter 7 of the above mentioned book. Samuelson's conservation law was also thoroughly analyzed there, and it was discovered to be the only conservation law with a von Neumann model. In addition, using the Noether theorem in standard neo-classical growth theory led to the discovery that, when consumption or the utility of consumption is maximized from the present into the infinite future, the "income-wealth ratio is constant" (for further details see Journal of Econometrics 1985; also found in The Selected Essay of Ryuzo Sato [Elgar, 1999], vol. II, chapter 18). This income-wealth conservation law was also derived by M. Weitzman, Kemp and others. But Weitzman's law was obtained using Bellman's principle of optimality not the general Noether theorem. Weitzman never mentioned the possibility that conservation laws other than the income-wealth law might also exist. In addition, unlike Sato (1985), he gave no answer to the question of whether the income-wealth conservation law would be found in a more realistic model such as one in which the discount rate changes over time.

As a matter of fact, the question of whether the income-wealth ratio is constant when the discount rate changes with time was first raised by Samuelson.

The author's analysis discovered the law that the rate of income to adjusted wealth is always constant when the discount rate is changing. This is the variable discount rate income-wealth conservation law.

The study of conservation laws had only just begun. Efforts were now made to discover new conservation laws in a form that was a further development of the Noether theorem. A representative example is the recent achievement on Nôno and Mimura. The latest achievements in this field were made public at a symposium held in April 2003 at New York University where papers by Samuelson, Weitzman, Mino, Hartwick, Cioppri, Russell-Cooper-Samuelson, and Sato were discussed. (For further details see Japan and the World Economy, vol. 16, no. 3 (august 2004), "Special Issue: Economic Conservation Laws and Optimizing Behavior," ed. T. Russell, pp. 234–415).

2. Survey of Chapters

Part One: Biased Technical Change

The Stability of the Solow-Swan Model with Biased Technical Change

21. Let me give an overview of each chapter in this book. In Chapter 2 it is shown that, by introducing endogenous technical change to the Solow-Swan model as discussed in part I of the introduction, a stable balanced growth path can be achieved under biased technical change, in short, under Sato-Beckmann-Rose factor augmenting technical change as well as under Harrod neutral technical change.

First, when the production function is given as a function of capital, labor and technical change, it can be written as follows

$$Y(t) = F[K(t), L(t), t] = F[A(t)\,K(t), B(t)\,L(t)] = \bar{Y}(t).$$

F is the linear and homogeneous neo-classical production function of K(t) and L(t). For simplification's sake, let us assume that B(t) and L(t) are exogenous. In the Solow-Swan model it is assumed that the capital increment = investment is determined by savings of only s% of Y, namely:

$$\dot{K}(t) = S(t) = sY(t), \text{ where } s = const,\ 0 < s < 1$$

If it is assumed that there is absolutely no technical change of capital, i.e. if A(t) = const = 1, i.e., Harrod neutral, then this economic system is able to achieve a stable, long-run growth path.

22. If the technical change rate of capital is exogenous in this Solow-Swan model, $A(t) = A_0e^{\alpha t}$, where α = const. > 0 capital measured by its efficiency, i.e. A(t) K(t) (defined as $\bar{K}(t)$), always grows faster at an annual rate of α% than K(t), which is determined endogenously by Y. In short, because the efficiency of capital

is higher than the efficiency of the economy as a whole, the shares of capital and labor cannot be constant, even in the infinitely long run t.

If the elasticity of factor substitution is greater or less than 1, the shares of capital and labor will both be zero. This means that a balanced growth path cannot be achieved.

23. In Chapter 2, A(t) is also assumed to be endogenously determined. This is a more realistic assumption in the macroeconomy. Moreover, both A(t) and K(t) are thought to be dependent on Y(t). As for the technical change rate of capital, it is assumed that a certain percentage of income will be allocated to technological development. In short, it is assumed that a nation will make appropriate investments in and allocations to both kinds of technical change to improve the quantity and quality of its physical capital. This means

$$\dot{\bar{K}}(t) = \frac{d(A(t)\,K(t))}{dt} = s\bar{Y}(t) = (s_1 + s_2)\bar{Y}(t)$$

Hence,

$$\frac{\dot{\bar{K}}(t)}{\dot{\bar{K}}(t)} = \frac{\dfrac{d\,(A(t)\,K(t))}{dt}}{A(t)\bar{K}(t)} = \frac{s\bar{Y}(t)}{\bar{K}(t)} = s_1\frac{\bar{Y}(t)}{\bar{K}(t)} + s_2\frac{\bar{Y}(t)}{\bar{K}(t)}$$

i.e.

$$\frac{\dot{A}(t)}{A(t)} + \frac{\dot{K}(t)}{K(t)} = s_1\frac{\bar{Y}(t)}{\bar{K}(t)} + s_2\frac{\bar{Y}(t)}{\bar{K}(t)}$$

The efficiency of capital rises only $\frac{\dot{A}}{A}$ and the amount of capital rises only $\frac{\dot{K}(t)}{K(t)}$. Moreover, is financed by the investment of s_1% of income $\bar{Y}$ of per unit capital (measured by efficiency), and $\frac{\dot{K}(t)}{K(t)}$ is created similarly by the investment of s_2%. The efficiency of labor B(t) as well is essentially unchanged, even though it assumed that it too is created in the same way by endogenous technical change. This is because labor L(t) is exogenous.

24. In long-run equilibrium, the shares of capital and of labor each preserves a constant value; as for the economy, $\bar{Y}(t)$ and $\bar{K}(t)$ continues to grow at the same rate, and $\frac{A(t)K(t)}{B(t)L(t)}$ slowly approaches a constant value.

Next, given the fact that return to capital is actually relatively stable and that interest rates and profit rates are also relatively stable, isn't there a contradiction in the model introduced here? The answer is that there is no contradiction. Why not? The reason is that the real economy is always on a path to long-run equilibrium but not at long-run equilibrium itself. This is the theory of endogenous technical growth shown in Sato [1963] where $\frac{\dot{B}(t)}{B(t)}$ = an average of 2.7% and $\frac{\dot{A}(t)}{A(t)}$ = an average of 0.7%; more over, it is explained as a path to a stable path. It is not in fact on a long-run balanced growth path itself, but merely on a path to reach that path.

A Model of Optimal Economic Growth and Endogenous Bias

25. Here, optimal technical change is introduced into the model described above. An analysis is made of how an economy can achieve balanced growth under biased technical change by combining optimal technical change and optimal growth.

The main themes here are the introduction of an investment function and a technical change function. In the case of the latter, moreover, technical change of capital and technical change of labor are treated separately. As an extension of the way of thinking in the preceding chapter, the technical change function with diminishing returns is assumed.

The model shows how a central planner can implement a plan that can most efficiently maximize welfare of per capita consumption by investing a certain amount of resources Y in three categories: increasing the amount of physical capital and raising both its technical change rate and that of labor.

26. Given the general factor-augmenting type (or biased type) of technical change under a neo-classical constant returns to scale production function, $Y = F[AK, BL]$, the society chooses three control variables m_1, m_2 and m_3 so as to maximize the discounted value of the sum of utility of consumption over an infinite time. Here rates of growth of efficiency of inputs K and L are given by technical progress functions,

$$\frac{\dot{A}(t)}{A(t)} = h_1\left(\frac{m_1 Y}{AK}\right)$$

$$\frac{\dot{B}(t)}{B(t)} = h_2\left(\frac{m_2 Y}{AK}\right)$$

It is also assumed that the transformation of income into capital goods (capital accumulation) is non-linear, with the function taking a form similar to that of the technical progress function:

$$\frac{\dot{K}(t)}{K(t)} = h_3\left(\frac{m_3 Y}{AK}\right)$$

Labor grows at a constant proportional rate,

$$\frac{\dot{L}(t)}{L(t)} = n.$$

Then the maximization problem can be written as

$$\underset{(m_1, m_2, m_3)}{\text{Max}} \int_0^\infty e^{-(\rho+n)t} F[AK, BL](1 - m_1 - m_2 - m_3)dt$$

Subject to

$$\frac{\dot{A}}{A}+\frac{\dot{K}}{K}=h_1(\cdot)+h_3(\cdot)$$

and

$$\frac{\dot{B}}{B}+\frac{\dot{L}}{L}=h_2(\cdot)+n.$$

There exists an interior solution such that, in a steady state, the effective capital growth rate is identical with that of effective labor i.e.

$$\frac{\dot{A}}{A}+\frac{\dot{K}}{K}=\frac{\dot{B}}{B}+\frac{\dot{L}}{L}.$$

27. This implies that Y, AK and BL are all growing at the same rate and the relative shares of capital and labor approach constant values. One of the paradoxes of the neo-classical model of growth is that it has a steady state only if technological progress is Harrod-neutral; a positive rate of labor augmenting has the same net consequences as a higher rate of growth of labor except that the per-capita output in steady state increases at the rate of increase in labor efficiency.

In contrast, the model presented in this chapter, has an interior solution and a steady state in which there is positive accumulation of capital and increases in efficiency of both capital and labor. A comparison of the two models seems to indicate that the crucial difference is in the assumption about the effect of expenditure investment and in increasing the efficiency of capital.

At steady state, the ratio of inputs measured in efficiency terms will remain constant and with it the factor shares. If one assumes a CEDD production function, then the elasticity of substitution is also a constant at steady state. However, before the system converges to the steady state, both factor shares and the elasticity of substitution will vary. Thus, it provides a theoretical growth model that is consistent with the empirical work of Sato (1970).

A Three Sector Model of Endogenous Hicksian Bias

28. This chapter uses the model of biased technical progress presented in the previous chapter and explores it in further detail. The economy is assumed to comprise of three sectors. The first sector produces a consumption good, the second produces a capital good while the third sector determines the innovation possibility frontier. The total quantity of labor at any time, L(t), is allocated among the three sectors:

$$L=L_C+L_K+L_R=L_0e^{nt},$$

Or, dividing by L,

$$1=l_{\mathrm{C}}+l_{\mathrm{K}}+l_{\mathrm{R}}. \tag{1}$$

The consumption good is produced using capital and labor. The constant-returns to scale production function is

$$C = F(AK, BL_C).$$

Where A and B are the efficiency factors of the inputs. Capital good is produced using only labor:

$$\frac{\dot{K}}{K} = \left(\frac{DL_K}{K}\right) - \delta = \left(\frac{Dl_k}{k}\right) - \delta \qquad \textbf{(2)}$$

where δ is the rate of depreciation, D is the efficiency of labor in the capital-good-producing sector, and $k = \frac{K}{L}$.

The rate of factor augmentation is determined in the research sector according to a concave innovation possibility frontier,

$$G(A, \dot{A}, B, \dot{B}, D, \dot{D}, E, \dot{E}, K, L_R) = 0 \qquad \textbf{(3)}$$

where E is the efficiency of labor in the research sector. As E is the only input into this sector, changes in E can be thought of as Harrod-neutral technical progress.

Under the assumption of factor-augmenting technical progress, capital enters the production function in efficiency terms, AK. An economy can choose between increasing the number of physical units of inputs or their efficiency, and the decision will be determined by marginal considerations. The question then is how these marginal conditions are affected by the formulation of the innovation possibilities frontier. The intertemporal optimization can now be stated as follows.

$$\max \int_0^\infty e^{-\rho t} \frac{1}{L} F(AK, BL_C) dt$$

subject to (1), (2) and (3).

29. The solution to the above problem will yield optimal bias and the derived bias in this model agrees with Sato (1970), which estimated that the growth in the efficiency of labor exceeds that of capital though both are positive. However, while Sato (1970) assumed that the economy produces one malleable good, in this model the economy is divided into three sectors with the assumption that the growth in productivity of labor in the capital good sector is less than that in the consumption goods sector.

If productivity growth is to be achieved only by the allocation of scarce resources to research, the question whether or not it is efficient to allocate some resources to increasing the efficiency of capital cannot be avoided. In microeconomic models, where the firm is the price-taker in the input market, the relative growth of input prices can be taken as the driving force. In macroeconomic models, the relative growth in input prices has to be derived from the model itself in terms of the intertemporal paths of resources and technology.

Formulating a model to address the question of optimal bias poses many problems. In Sato *et al.* (1999) and in this chapter, an attempt has been made to develop tractable models to analyze the optimal intertemporal allocation of resources leading to a steady-state bias. These papers reveal the crucial role of the technology of the production and research sectors in determining the allocation of resources for physical capital accumulation and for the optimal innovation frontier.

Estimation of Biased Technical Progress

30. Empirical works on neo-classical growth models led to the recognition that technological progress is the dominant factor in the growth of *per capita* income. There are four questions: (1) How important is technological and technical progress in the process of economic growth? (2) What is the cause of technical progress -is it exogenous or endogenous to the system? (3) How is technological change transmitted into technical progress in the macroeconomy? and (4) If technical progress can be classified as labor saving, neutral or capital saving, is there any systematic bias in an economy towards any particular kind of technical progress and, if so, why?

This chapter gives a partial answer to the above questions. By using the estimation equations for the growth of efficiencies of capital and labor (see Sato [1970]), we compare the growth of factor productivities in the United States and Japan. The growth rate of capital's efficiency is estimated from

$$\frac{\dot{A}(t)}{A(t)} = \frac{\sigma\frac{\dot{r}}{r} - \frac{\dot{Y}}{Y} + \frac{\dot{K}}{K}}{\sigma - 1}, \ \sigma \neq 1$$

and that of labor's, from

$$\frac{\dot{B}(t)}{B(t)} = \frac{\sigma\frac{\dot{w}}{w} - \frac{\dot{Y}}{Y} + \frac{\dot{L}}{L}}{\sigma - 1}, \ \sigma \neq 1$$

where $\sigma =$ the elasticity of factor substituting, $\frac{\dot{r}}{r} =$ growth rate of returns to capital and $\frac{\dot{w}}{w} =$ the growth rate of the wage rate.

Because of the impossibility theorem of Diamond-McFadden-Sato, we estimate average elasticity of substitution assuming Hicks neutral technical progress. Using this value, we then estimate the average growth rates of efficiencies of capital and labor. But the estimate of the elasticity based on the assumption of Hicks neutrality has a bias if technical progress is indeed biased. To test the robustness of the results, we estimate the growth rates for different elasticities of substitution.

31. For the U.S. data (1909–1993), what is striking about various estimates is that the growth rate of the efficiency of labor is consistently four to five times that of capital, i.e.

$$\frac{\dot{A}}{A} = 0.004 \textit{ and } \frac{\dot{B}}{B} = 0.019 \sim 0.020$$

The same method is used to estimate the growth rates in efficiencies in Japan. There is considerable variation in the relative magnitudes of the growth rates of the two efficiencies in Japan. In fact, for elasticities of substitution greater than 0.8, $\frac{\dot{A}}{A}$ is positive and greater than $\frac{\dot{B}}{B}$ while for lower values of elasticity, $\frac{\dot{A}}{A}$ is not only less than $\frac{\dot{B}}{B}$, but turns negative.

This is in direct contrast to the United States where the growth rate of A(t) decreases as elasticity increases. This can be explained by inefficiency of the capital market in Japan, even before the collapse of the bubble economy. There were hidden problems of over investment and non-performing debt in the Japanese economy.

A Note on Modelling Endogenous Growth

32. This chapter explores the conditions for the existence of a steady state in an optimal growth model with factor-augmenting technical progress. Given a non-linear "investment" function, the consumption function is;

$$C = C\,(A, K, B, L;\ \dot{K}, \dot{L}, \dot{A}, \dot{B})$$

where A and B = efficiencies of capital K and labor L respectively.

There are several sufficient conditions under which the "effective" capital-" effective" labor ratio can attain a constant value in steady state.

They are;

(1) $\lambda^2 C = C\,[\lambda A, \lambda K, \lambda B, \lambda L; \lambda(A\dot{K}), \lambda(B\dot{L})]$

(2) $\lambda^2 C = C\,[\lambda A, \lambda K, \lambda B, \lambda L; \lambda(\dot{A}K), \lambda(\dot{K}A), \lambda(\dot{B}L), \lambda(\dot{L}B), \lambda(\dot{A}\dot{K}), \lambda(\dot{B}\dot{L})]$

(3) $\lambda^2 C = C\left[\lambda A, \lambda K, \lambda B, \lambda L; \lambda(AK)\,\phi_1\left(\frac{\dot{A}}{A}\right), \lambda(AK)\,\phi_2\left(\frac{\dot{K}}{K}\right),\right.$
$\left.\lambda(BL)\,\phi_3\left(\frac{\dot{B}}{B}\right), \lambda(BL)\,\phi_4\left(\frac{\dot{L}}{L}\right)\right]$

Examples which satisfy each one of the above conditions are presented.

A special case of (3) is the additive "investment" function used in Chapter 3 of this book.

The general form is,

$$C = F\,[AK, BL] - [M_1 + M_2 + M_3]$$

or

$$C = F\,[AK, BL] - \left[AK\phi_1\left(\frac{\dot{A}}{A}\right) + AK\phi_2\left(\frac{\dot{K}}{K}\right) + BL\phi_3\frac{\dot{B}}{B}\right]$$

where M_i = the amount of investment for each category.

This implies that the "investment" functions for technical change and capital accumulation take the form;

$$\frac{\dot{A}}{A} = \phi_1^{-1}\left(\frac{M_1}{AK}\right)$$
$$\frac{\dot{K}}{K} = \phi_2^{-1}\left(\frac{M_2}{AK}\right).$$
$$\frac{\dot{B}}{B} = \phi_3^{-1}\left(\frac{M_3}{BL}\right)$$

These are exactly the functions used in chapter 3.

Finally, this chapter shows that an optimal growth path under endogenous bias is stable even under more general "investment" functions than those used in chapter 3 of this book.

Technical Change and International Competition

33. This chapter is not directly related to biased technical change but is closely related to technical change resulting from R&D activities in basic and applied technologies.

A differential game model is used to study the basic nature of international competition between the technology leaders and latecomers. In this model, two monopolistic firms in the two types of countries engage in R&D activities to produce similar products and export them in the world market. The technologically advanced firm in country A engages in both basic and applied research, whereas the firm in country B imports basic technology from country A and improves on it.

There are three relevant parameters, which will affect the outcome of the competition: (1) the index of diffusion of basic technology (2) the index of relative efficiency of applied technology, and (3) the index of cost of sharing of basic research.

It is shown that the final outcome (i.e. who wins the competition?), depends on the strategies the two firms employ: an open-loop or a closed-loop strategy. Under normal conditions, the open-loop strategy may impart equal market shares among the world competition. A paradoxical case may occur when the less technologically advanced country increases the cost of sharing of basic technology. The country with no domestically produced basic research, can now win even if its efficiency in applied technology is lower than that of the country producing basic research. The closed-loop strategy may eliminate the paradox.

Part Two: Conservation Laws

Optimal Economic Growth: Test of Income/Wealth Conservation Laws in OECD Countries

34. This chapter attempts to derive several economic conservation laws and to test the validity of the optimal growth models using the income/wealth ratios for the United States, Japan, and other OECD countries.

In this traditional approach, one major concern arises from determining the nature of appropriate choices in investment and how these decisions will influence the economy's path of long-run development. This method of analysis involves typical application of "optimal" growth theory. Another application of the optimal growth theory occurs when economists and statisticians attempt to justify and estimate the single index of "national income" from a theoretical point of view. However, little empirical work has been done to test the validity of the second aspect of the growth theory, namely the testing of an appropriate measure of national income.

The purpose of this chapter is to unify both aspects of the optimal growth theory with a general theory of "economic conservation laws". The theory of conservation law involves the identification and discovery of hidden invariant quantities in a dynamic system. In this dynamic economic system, as in a dynamic physical system, it is suspected that a certain variable remains unchanged during its process of evolution, as long as the system follows an optimal trajectory. In a growing economy, the variable that is invariant is called the conservation law. By uncovering the existence of conservation laws and by formulating the operational concepts associated with them, one can test the valdity of both the optimal growht models as well as the measurement of national income. It has been shown that in an optimally controlled ecomomy, the ratio of income to wealth remains invariant. This is the income/wealth conservation law.

35. The model formulating this conservation law is presented. It is assumed that the consumption of the economy depends on output, which in turn depends on a vector of capital goods, a vector of investment and labor input, so that

$$C = C[Y(K;\ \dot{K};\ L;\ \dot{L})], \tag{1}$$

where Y = output, which depends on $K = (K_1, \ldots K_n) = n$ capital goods, $\dot{K} = (\dot{K}_1, \ldots \dot{K}_n) = dK/dt =$ investment and L labor input, exogenously given by $L = \lambda L$. Without loss of generality, we can assume that $\frac{dC}{dY} = 1$ and (1) can be written as

$$C = F(K;\ \dot{K};\ L;\ \dot{L})$$

Consumption per capita is given by

$$\begin{aligned} c = \frac{C}{L} &= F\left(\frac{K_1}{L}, \ldots, \frac{K_n}{L}; \frac{\dot{K}_1}{L}, \ldots, \frac{\dot{K}_n}{L}; 1; \frac{\dot{L}}{L}\right) \\ &= F(k_1, \ldots, k_n;\ \dot{k}_1 + \lambda k_1, \ldots, \dot{k}_n + \lambda k_n;\ 1;\ \lambda \end{aligned}$$

where $k_i = \frac{K_i}{L}$ and $\dot{k}_i = \left(\frac{\dot{K}_i}{K_i} - \lambda\right) k_i \quad i = 1, \ldots, n.$

Thus we have

$$c = f(k;\ \dot{k} + \lambda k;\ \lambda).$$

The society's objective is to maximize the discounted future value of consumption per capita, c(t) = f(t), as

$$J = \int_0^\infty e^{-\rho t} f(k; \dot{k} + \lambda k); \lambda) dt \rightarrow \text{Max}$$

Using the Noether theorem and its invariance principle, the general expression for the conservation law can be derived as

$$\rho = \frac{\textit{Income per capita}}{\textit{Wealth per capita}}$$

36. Conservation laws vary with the type of objective function depcted, such as the maximization of the aggregate consumption of the maximization of per-capita consumption. The operational concept of "wealth-like quantity" is identified, although the Goldsmith-Kendrick standard definition of "net national wealth" should not always be used. The last section of this chapter takes up an emprical analysis to determine how different economies have achieved long-term (optimal) growth. The U.S. economy has been operating rather efficiency, whereas the Japanese economy, after the oil shocks of the 1970's, has behaved differently, leading into the bubble period of the early 1990's.

Economic Conservation Laws as Indices of Corporate Performance

37. This chapter attempts to apply economic conservatio laws to corporate behavior at the microeconomic level. At the macro level, efficient operation and performance of nations are analyzed using the income-wealth conservation law. Corporate management needs to be able to judge whether or nor a company can achieve the maximum long-run profit.

In modern finance and accounting theory several different criteria are considered for measuring corporate performance. Investors in stocks, for exmaple, will use P-E ratios etc. to judge how high or low a company's stock price is relative to it's profitability. Corporate mangers will analyze management conditions collectively by considering whether or not profitability per unit capital is at a satisfactory level, or by looking at trends in productivity. In the final analysis, the yardstick for successful corporate management is whether corporate profits are growing at a satisfactory level.

Purely from an economic perspective, a corporation may be analyzed on the assumption that is operating to maximize profits over a certain period under the given conditions. In actually, however, production costs and R&D activity etc. at such an institution may be unpredictable. Since the numerical values of crucial parameters are ambiguous, it is impossible to judge whether the company's reported profits are being maximized or not. Under such circumstances, managers regard positive growth of profits as the criterion of succss. In short, managers are regarded as successful as long as profits are growing even though profits may not be maximized.

Hence corporate management needs to be able to judge whether or not under the given conditions a company is being managed in a way that will maximize long-term profit growth. This chapter aims to apply macroeconomic methods to microeconomic organizations and to come up with an assessment of corporate performance by applying economic conservation laws to individual companies and individual industries.

38. Because theoretical analyses using economic conservation laws are not yet in general use, the paper begins with a general model and an explanation of the Noether theorem for deriving conservation laws. The forms that corporate profit maximization take often idffer greatly depending on the nature of management and the management variables that are emphasized. In order to analyze these forms, several representative corporate behaviors found in the existing literature are presented as examples. The view of long-term profit maximization shared in these examples is explored and an attempt is made to verify the workings of their profit maximization behavior using empirical data. As a theoretical assumption, the important criterion is whether or not the discount rate is fixed. Economic conservation laws under different hypotheses are then derived from this.

To explain Noether's invariance principle, a profit equation of the firm π, which is twice continuously differentiable in each of its $2n + 1$ arguments, is considered.

We have the firm's long-run profit,

$$J(x) = \int_a^b \pi(t, x(t), \dot{x}(t))\, dt$$

where x is the vector of all quantities and prices, which the firm controls to maximize the long-run profit

$$J(x),\ x(t) = (x^1(t), \ldots x^n(t)),\ and\ \dot{x}(t) = (\dot{x}^1(t), \ldots, \dot{x}^n(t)).$$

The firm's problem is to maximize the long-run profit

$$J(x) \rightarrow Max.$$

The following specialized examples of corporate behavior are taken from the literature.

1. Capital investment model

If P(k) is the profit rate generated with capital k producing output f(k), revenue is given by pf(k) = P(k). Let C(I) be cost of investment and b the constant proportional decay of capital stock, then $C(I) = C(\dot{k} + bk)$. The firm's objective is to maximize the present value of the net stream over time,

$$\Pi = P(k) - C(\dot{k} + bk),\ i.e.$$

$$\max \int_0^\infty e^{-\rho t}[P(k) - C(\dot{k} + bk)]\, dt$$

subject to the initial conditions.

2. Price-quantity adjustment model

In an oligopolistic market profit depends on price p, quantity q and $\dot{q}$, rate of change of q;

$$\Pi = (p, q, \dot{q})$$

but since other producers participate, $\dot{p} = f(p, q)$

Thus, the firm's objective is to maximize

$$\max \int_0^\infty e^{-\rho t}\pi(p, q, \dot{q})\, dt$$

Subject to

$$\dot{p} = f(p, q) \textit{ and the initial conditions.}$$

3. R&D profit maximization model

The typical firm in oligopolistic competition attempts to increase its profits by adjusting output and adopting cost-reducing innovations. To do this, the firm engages in both basic and applied research. The firm's objective is to maximize

$$J = \int_0^\infty e^{-\rho t}[R(p, y) - C(A, \dot{A}, B) - T(B, \dot{B})]\, dt$$

where p = price of output y, A = level of applied research and B = level of basic research.

Since there exists no stylized facts as to which example fits the observed data, we will use the general case of maximizing

$$J(x) = \int_a^b \pi(t, x(t), \dot{x}(t))\, dt.$$

Using the Noether theorem invariance conditions, we derive several conservation laws. There are six conservation laws that have been discovered including the one used to test macroeconomic performance in the previous chapter.

This conservation law states that the current value of profit and investment must be equal to the discount rate multiplied by the discounted value of the firm. We will call this the *Total Value Conservation Law of the Firm.*

39. Thomas Mitchell's comments are available at the end of the chapter as an appendix.

Empirical Tests of the Total Value Conservation Law of the Firm

40. The previous chapter shows how economic conservation laws could be applied in the evaluation of corporate behavior. This chapter applies a conservation law to individual industries and companies, and empirical tests are presented to show the possibility that there are ways to evaluate manager behavior relative to the profit-maximization goal. The results represent the first extensive investigation to apply a conservation law using Japanese corporate data.

Within the context of the *total value conservation law of the firm* derived in the previous chapter, empirical evidence using the Japanese corporate data is examined. In conducting the empirical analysis, a constant discount rate is assumed. We can test whether the conservation laws hold for firms by using data from their financial statements and market prices of their securities and shares. Because the theory predicts that if a firm optimizes profits, the derived ratio of income creation to the value of the firm should be relatively stable, the volatility of this ratio (σ_β) should be negatively correlated with some good measure of performance of the firm.

We then perform a regression analysis using the performance indicator of the firm (*PI*) and its level of leverage (*L/V*), as predictors of the volatility of its ratio, σ_β. Testing the model with the Nikkei Performance Index as *PI*, we obtain a statistically significant negative coefficient for *PI*. Using return on assets as an alternative performance index on the same selection of firms used previously, yields a statistically significant positive correlation coefficient for *PI*. Testing the model with another measure of performance, the growth of total assets over the sample period, indicates a statistically significant positive correlation for *PI*.

Thus the regression results are inconclusive. Using the Nikkei index shows the possibility that better managed firms are related to low volatility of the ratio, which is consistent with our assumption of a constant discount rate. However, indices constructed from financial statement data suggest, that greater the values of either return on assets or the growth of total assets, the higher the volatility of the ratio.

Another data issue is survivorship bias. Since a long history of company data is required for the regression analysis, companies that are short-lived because of poor management or any reasons are excluded from the sample data. Thus the tests in this chapter are not ideal for identifying specific firms with profit maximizing operations.

41. This chapter attempts to abstract the real world and construct a convincing model as a good approximation of firm behavior. In traditional valuation theory under uncertainty, firms are given their discount rate from the capital markets, as opportunity cost of their investment. Capital markets force managers to optimize and surviving firms are assumed to be optimizing. On the other hand, given the relevant value of the discount rate, it is important to examine whether a firm has been operating optimally even under the assumption of certainty. From this perspective, we are interested in longer-term trends rather than daily market fluctuations in firm valuation.

In this regard, we have shown that the assumption of a constant discount rate and profit-maximizing behavior is consistent with observed data for many top-performing Japanese firms during the period of 1980 to 2002. Although the results are tentative because of the limited data period and data availability, they suggest that it is promising to extend the notion of conservation laws to the industry or firm level.

Hartwick's Rule and Economic Conservation Laws

42. By using the theory of conservation laws we derived two investment rules for optimal resource extraction and intergenerational equity: first, the so-called Hartwick rule as a special case of a more general rule; and second, a new investment rule for Benthamite utility maximization.

We start with a simple model of optimal capital accumulation with extraction cost of resources. The aggregate income is spent by consumption expenditure, investment on physical capital, and extraction of natural resources. The production function is subject to the condition of constant returns to scale.

43. The society's problem is to find the largest constant consumption with a feasible pattern of resource use over the infinite time horizon. Hartwick [1977] derived a rule that investment must be equal to the rents from exhaustible resources in order to maintain a path of constant consumption. We examine whether Hartwick's rule can be derived from one of the conservation laws and whether there is another policy rule. It is shown that, in fact, Hartwick's rule is a special case when the conservation law (H = Hamiltonian) takes a special value, i.e. $H = 0$. There are many other rules when $H \neq 0$.

Next, we study the Hotelling differential equation. The solution of the differential equation specifies how the capital-resource ratio must follow the optimal path. A complete closed form solution is given for the Cobb-Douglas production function.

44. We then present a simple model of optimal resource extraction under the Benthamite utility function. The economy has a representative agent with a well-defined utility function. Then the economy's objective is to solve the Benthamite problem of dynamic optimization with a positive discount rate.

By using the theory of conservation laws, we derive the income-wealth invariance law which yields a new investment rule. The rule depends not only on the discount rate but also on how profit changes over time. The Benthamite investment rule coincides with Hartwick's rule, if and only if the discount rate is zero

Appendix

45. Excerpt from Ryuzo Sato [1999], Chapter 7 Dynamic Symmetries and Economic Conservation Laws (Sections I, II & III) is appended for those who are not familiar with the methodology of Lie group application.

References

Hartwick, J. [1977], "Intergenerational Equity and the Investing of Rents from Exhaustible Resources," *American Economic Review*, 66, 972-974.

Hicks, John [1932, 1963], *The Theory of Wages*, 2nd Ed. (1963), Macmillan.

Kendrick, John and Sato, Ryuzo [1963], "Factor Prices, Productivity, and Economic Growth," *American Economic Review*, 53, 974-1003.

Keynes, John M. [1936], The General Theory of Employment, Interest and Money, Harcourt, Brace and Company, Inc.

Nôno, T. and Mimura, F. [1975, 1976 & 1977], "Dynamic Symmetries, I, II & III," *Bulletin of Fukuoka University of Education.*

Ramsey, Frank [1928], "A Mathematical Theory of Saving," *Economic Journal*, 38, 543-559.

Rose, Hugh [1969], "The Condition for Factor Augmenting Technical Change," *Economic Journal*, 78, 966-971.

Russell, Thomas [2004], "Special Issue: Economic Conservation Laws and Optimizing Behavior," T. Russell (ed.), *Japan and The World Economy*, Vol. 16 (3), 234-415.

Samuelson, Paul [1965], "A Theory of Induced Innovation along Kennedy-Weizsäcker Lines," *Review of Economics and Statistics*, 47, 343-356.

Samuelson, Paul [1970], "Law of Conservation of the Capital-Output Ratio," *Proceedings of the National Academy of Sciences, Applied Mathematical Science*, 67, 1477-1479.

Sato, Ryuzo [1980], "The Impact of Technical Change on the Holotheticity of Production Function," *Review of Economic Studies*, 47, 767-776.

Sato, Ryuzo [1981, 1999], *Theory of Technical and Economic Invariance: Application of Lie Groups*, Academic Press, 1981; Revised Edition, Elgar, 1999.

Sato, Ryuzo [1996], *Growth Theory and Technical Change: The Selected Essays of Ryuzo Sato*, Volume One, Elgar.

Sato, Ryuzo [1999], *Production, Stability and Dynamic Symmetry, The Selected Essays of Ryuzo Sato*, Volume Two, Elgar.

Sato, Ryuzo [1970], "The Estimation of Biased Technical Progress and the Production Function," *International Economic Review*, 11, 179-208.

Sato, Ryuzo and Beckmann, M. [1968], "Neutral Inventions and Production Functions," *Review of Economic Studies*, 35, 57-65.

Solow, Robert [1956], "A Contribution to the Theory of Economic Growth," *Quarterly Journal of Economics*, 70, 65-94.

Solow, Robert [1957], "Technical Change and the Aggregate Production Function," *Review of Economics and Statistics*, 39, 312-320.

Swan, T. W., "Economic Growth and Capital Accumulation," *Economic Record*, 32, 334-361.

Part One

BIASED TECHNICAL CHANGE

Chapter 2

THE STABILITY OF THE SOLOW-SWAN MODEL WITH BIASED TECHNICAL CHANGE

1. Introduction

1. In the traditional Solow-Swan model of economic growth, the economy can achieve balanced growth if technical change is Harrod neutral, that is to say, if technical change is labor-augmenting (Russell [2004]). But the system *cannot* achieve balanced growth under either the Hicksian or the general factor-augmenting type of technical change, where the efficiency of capital is improving. This is because the output-capital ratio *cannot* attain a constant value even in the long run. The ratio is ever increasing, which implies that the income share of capital (or of labor) will eventually approach the extreme value of either zero or unity, depending on the underlying technology, or to be more specific, depending on the elasticity of factor substitution. Stable balanced growth does *not* exist.

This well-known conclusion hinges on the crucial assumption that technical change is exogenous. Given an exogenous growth rate of labor, say n% a year, an exogenous type of labor-augmenting efficiency improvement of β% is not a complicating factor. The system will simply adjust to the $(n + \beta)$ growth of the exogenous force when there is technical progress. Thus, the growth rate of labor measured in its efficiency units is the sum of the physical growth rate of labor and its efficiency growth rate.

On the capital side, however, the situation is completely different. Capital is endogenously determined by the savings and investment behaviour of the economy. Under the traditional Solow- Swan model, a fraction of income (or output) is saved and invested. If the efficiency improvement of capital is exogenous, capital measured in its efficiency units depends not only on an endogenous force but also on an exogenous factor. This creates a problem in that the output-capital ratio depends on both endogenous and exogenous factors. This ratio is ever increasing as long as there is a positive exogenous efficiency improvement in capital. Output will always increase faster than physical capital, and the output-capital ratio can never achieve a constant value.

2. In Sato-Ramachandran-Lian (1999), it is shown that if the rate of technical progress is optimally controlled, the system can achieve balanced growth even under the general factor-augmenting type of technical change. Optimally controlled technical change is, by definition, endogenous. Hence, the endogeneity assumption is the critical element for stability.

The central planning authority will be able to employ a set of investment policies. There are three choices: (1) to invest in physical capital, (2) to invest to improve the efficiency of capital, and (3) to invest to improve the efficiency of labor. The optimal combination of all three types of investment must be made to maximize the social welfare of a nation in the long run. This is an extension of the traditional optimal savings problem, but, unlike the traditional analysis, the optimal amount saved is not automatically invested in physical capital. Instead, it will be invested in three different types of welfare-enhancing projects, namely, in physical capital and in improvements to both capital and labor efficiencies.

Stability conditions depend on the properties of the investment functions. If the investment functions have standard concavity conditions an optimally controlled economy can achieve stability in the saddle point sense.

3. In this analysis, we want to show that balanced growth is achievable even without the optimality conditions. In other words, even under the Solow-Swan model of fixed savings ratios, the general factor-augmenting type of technical change will impose no additional restrictions on stability conditions. This is because fixed savings ratios satisfy the (weak) concavity condition.

As long as capital-augmenting technical change is generated within the economic system, i.e. as long as it is endogenous, the system is stable. The failure to achieve balanced growth equilibrium under the factor-augmenting type of technical change is due to the fact that it is assumed to be exogenous.

2. The Model

4. Let aggregate output at time t, $Y(t)$, be produced by capital K(t) and Labor L(t) under a constant-returns-to-scale neo-classical production function F,

$$Y(t) = F[K(t), L(t)]$$

Let technical change be the factor-augmenting type with $A(t)$ and $B(t)$ as the efficiency improvement factors of $K(t)$ and $L(t)$ respectively. Then, aggregate output after technical change is represented by the transformed production function,

$$\bar{Y}(t) = F[\bar{K}(t), \bar{L}(t)] \tag{1}$$

where $\bar{Y}(t) =$ output after technical change and $\bar{K}(t) = A(t) \cdot Kh$ and $\bar{L}(t) = \bar{B}(t) \cdot L$ are the factor-augmenting type. Equation (1) can be viewed as the production function after K has been transformed to $\bar{K}$ and L to $\bar{L}$.

5. To begin with the simplest case, it is assumed that labor and its efficiency factor are both exogenously given as,

$$\frac{\dot{\bar{L}}}{\bar{L}} = \frac{(BL)}{BL)} = \frac{\dot{B}}{B} + \frac{\dot{L}}{L} = \beta + n = \lambda > 0 \tag{2}$$

where $\bar{L} = B.L$ = labor measured in its efficiency unit or "effective" labor, β = growth rate of labor-augmenting technical change, and n = the growth rate of labor.

Let "effective" capital, or capital measured in its efficiency unit, be given by $\bar{K} = A.K$. Assume that $\bar{K}$ is endogenously determined by

$$\dot{\bar{K}} = (AK) = s\bar{Y},\ 1 > s > 0 \tag{3}$$

where s = a fraction of output (or income) Y used to create additional effective capital. The fraction s may be divided into the fractions s_1 and s_2, where s_1 is devoted to improve the efficiency of A and s_2 is used for (physical) capital accumulation, i.e.,

$$\dot{\bar{K}} = (AK) = (s_1 + s_2)\bar{Y} = s_1\bar{Y} + s_2\bar{Y} -,\ s_i > 0 \text{ and } s_1 + s_2 = s \tag{4}$$

By dividing the above by $\bar{K}$, equation (4) may be expressed as

$$\begin{aligned} \frac{\dot{\bar{K}}}{\bar{K}} &= \frac{(AK\ )}{(AK)} = \frac{\dot{A}}{A} + \frac{\dot{K}}{K} = \frac{s_1\bar{Y}}{AK} + \frac{s_2\bar{Y}}{AK} \\ \frac{\dot{\bar{K}}}{\bar{K}} &= \frac{\dot{A}}{A} + \frac{\dot{K}}{K} = s_1\frac{\bar{Y}}{AK} + s_2\frac{\bar{Y}}{AK} \end{aligned} \tag{5}$$

Equation (5) is comparable to equation (2). Capital measured in its efficiency unit depends on the growth rate of efficiency factor $s_1\frac{\bar{Y}}{\bar{K}}$ and the growth rate of physical capital $s_2\frac{\bar{Y}}{\bar{K}}$. The difference between (2) and (5) is that in (2) the growth rate of labor measured in its efficiency unit is exogenous, whereas, in (5) the growth rate of capital measured in its efficiency unit is endogenous. The growth rate of effective capital depends on the (effective) output – (effective) capital ratio, $\frac{\bar{Y}}{\bar{K}}$. Hence, (5) can be written as,

$$\frac{\dot{\bar{K}}}{\bar{K}} = s\frac{\bar{Y}}{\bar{K}} \tag{5'}$$

or

$$\begin{aligned} \frac{\dot{A}}{A} &= s_1\frac{\bar{Y}}{\bar{K}} \\ \frac{\dot{K}}{A} &= s_2\frac{\bar{Y}}{\bar{K}},\quad s = s_1 + s_2 \end{aligned} \tag{5''}$$

3. Stability

6. Define the capital-labor ratio in efficiency unit as

$$\bar{k}(t) = \frac{\bar{K}(t)}{\bar{L}(t)} = \frac{A(t)\cdot K(t)}{B(t)\cdot L(t)},$$

Then

$$\frac{\dot{\bar{k}}}{\bar{k}} = \frac{\dot{\bar{K}}}{\bar{K}} - \frac{\dot{\bar{L}}}{\bar{L}} = s\frac{\bar{Y}}{\bar{K}} - \lambda = sF\left(1, \frac{\bar{L}}{\bar{K}}\right) - \lambda$$
$$= sF\left(1, \frac{1}{\bar{K}}\right) - \lambda$$
$$= sf\left(\frac{1}{\bar{K}}\right) - \lambda \qquad \textbf{(6)}$$

A growth path with factor-augmenting technical change is stable if,

$$\frac{d\left(\frac{\dot{\bar{k}}}{\bar{k}}\right)}{d\bar{k}} = s\frac{df}{d\bar{k}} = s(-1)\frac{df}{d\left(\frac{1}{\bar{k}}\right)}\bar{k}^{-2} < 0$$

Since $\frac{df}{d(\frac{1}{\bar{k}})}$ = the marginal product of effective labor is always positive, the above is always satisfied. Thus, a balanced path is stable under the endogenous factor-augmenting type of technical change.

7. Whenever $\frac{\dot{\bar{k}}}{\bar{k}}$ is positive, $\bar{k}$ must be increasing, whereas whenever $\frac{\dot{\bar{k}}}{\bar{k}}$ is negative it must be decreasing. When $k = k^*$, $\frac{\dot{\bar{k}}}{\bar{k}} = 0$, which implies that $sf\left(\frac{1}{\bar{k}}\right)$ must be

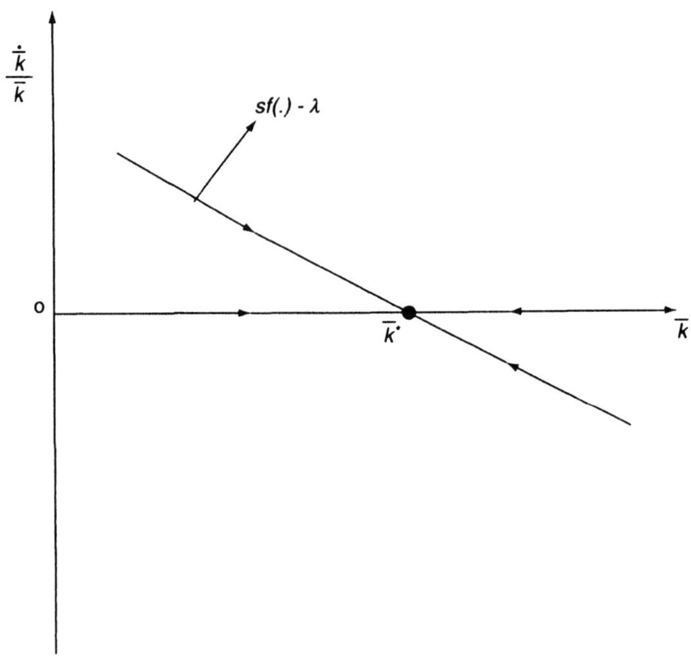

Figure 1. Stability of the System.

exactly equal to 1 (see Figure). At $\bar{k} = \bar{k}^*$, the growth rate of effective capital is identical to the growth rate of effective labor. The (effective) output-(effective) capital ratio remains constant.

4. Endogenous Labor-Augmenting Technical Change

8. Suppose that the efficiency improvement of labor also depends on the amount of expenditure devoted to education and training. Then equation (2) must be modified to

$$\frac{\dot{\bar{L}}}{\bar{L}} = \frac{(\dot{BL})}{(BL)} = \frac{\dot{B}}{B} + \frac{\dot{L}}{L} = s_3 \frac{\bar{Y}}{\bar{L}} + n \tag{1}$$

Labor efficiency now depends on the amount of money spent for improvement purposes. It is determined by a fraction (s_3) of income per labor measured in its efficiency unit. This is certainly more realistic than the case of exogenous technical change. With this endogeneity assumption, equation (6) must be changed to

$$\frac{\dot{\bar{k}}}{\bar{k}} = \frac{\dot{\bar{K}}}{\bar{K}} - \frac{\dot{\bar{L}}}{\bar{L}} = s\frac{\bar{Y}}{\bar{K}} - s_3\frac{\bar{Y}}{\bar{L}} - n, = sf\left(1, \frac{1}{\bar{k}}\right) - s_3F(\bar{k}, 1) - n, 1 >; s + s_3 > 0.$$

Letting

$$F\left(1, \frac{1}{\bar{k}}\right) = f\left(1, \frac{1}{\bar{k}}\right) \textit{ and } F(\bar{k}, 1) = g(\bar{k})$$

the above becomes

$$\frac{\dot{\bar{k}}}{\bar{k}} = sf\left(\frac{1}{\bar{k}}\right) - s_3 g(\bar{k}) - n \tag{9}$$

The stability condition is satisfied if,

$$\frac{d\left(\frac{\dot{\bar{k}}}{\bar{k}}\right)}{d\bar{k}} = s\frac{df}{d\bar{k}} - s_3\frac{dg}{d\bar{k}} <, 0$$

$$s = s_1 + s_2$$

$$0 < s_1 + s_2 + s_3 < 1 \tag{10}$$

The first term of the above, $s\frac{df}{d\bar{k}}$, is shown to be negative (see equation (7)). The second term is always positive, for $\frac{dg}{d\bar{k}}$ is the marginal product of effective capital. Hence equation (10) is automatically satisfied. In fact, by introducing endogenous labor-augmentation, the system becomes more stable.

5. Concluding Remarks

9. In the present analysis, it was shown that in the traditional Solow-Swan model of neo-classical growth, the economy can achieve stability as long as the efficiency improvement of capital is associated with the amount of expenditure devoted to quality improvement. If the axiom "there is no such thing as a free lunch" is true, it is more realistic to assume that efficiency improvement is the result of investment. There may be externalities in the world economy such that one nation may be able to obtain efficiency improvements in both capital and labor not as a result of its own investment but through the investment of some other nation. But this is not usually the case.

10. The result of the present analysis is not surprising. The efficiency factor of capital A(t) or labor B(t) is treated as another form of "capital" growing independently of the original capital K(t) itself. Since the efficiency factor A(t) is multiplicative to K and is growing at $s_1 \frac{\bar{Y}}{\bar{K}}\%$ a year, the growth rate of $K = A \cdot K$ is the sum of the two components, i.e. $\frac{\dot{A}}{A}$ and $\frac{\dot{K}}{K}$.

Empirical justification for this type of modeling can be found in many research works. For instance, the paper by Sato, Ramachandran and Kim [1999] has shown that the average growth rate of A is less than that of B both in the United States and Japan:

$$\frac{\dot{A}}{A} = 0.004 \text{ and } \frac{\dot{B}}{B} = 0.020 \text{ in the U.S.A; and } \frac{\dot{A}}{A} = 0.001 \text{ and } \frac{\dot{B}}{B} = 0.050 \text{ in Japan.}$$

This implies that the original Kaldorian description of the state of a mature Economy is still valid both in theory and in practice. Thus, the return to capital is basically trendless, but the wage rate is increasing, while the capital-labor ratio (in efficiency units) is relatively stable. This makes the relative income shares of capital and labor also relatively constant over the long run under the general factor-augmenting type of technical change. The above conclusion need not depend on the traditional Harrod neutrality assumption.

References

Russell, Thomas [2004], "On Harrod Neutral Technical Change." Mimeo, December 2004.

Sato, R., Ramachandran, R., and Kim, Y. [1999], "Estimation of Biased Technical Progress" in *Global Competition and Integration*, ed. R. Sato, R. Ramachandran, and K. Mino, Kluwer.

Sato, R., Ramachandran, R., and Lian, C. [1999], "A Model of Optimal Economic Growth and Endogenous Bias." *Macroeconomic Dynamics*. Vol. 3.

Chapter 3

A MODEL OF OPTIMAL ECONOMIC GROWTH WITH ENDOGENOUS BIAS

1. Introduction

Economists have always accorded technological change a crucial role in the interpretation of long-term trends. Theoretical considerations have shown that a clear distinction has to be made between movements along the production function due to changing relative scarcity of inputs and movements of the function itself due to technological change. Further research has been aimed at explaining the rate of shift of the function, the continued growth of wages in the absence of any persistent trend for interest rate, and the constancy of the factor shares.[1] The resulting discussion, instead of leading to a commonly accepted paradigm, has generated three distinct trends in the literature.

The first of these can be traced to Hicks' (1963) *Theory of Wages* in which he applied the analysis of neoclasical theory to the examination of the factor incomes without assuming, as in some of the earlier works, homogenous output and inputs. In the process of developing the analysis, he argued that technological change will be biased toward saving the input whose relative price was increasing; this bias and changes in the elasticity of substitution were used to explain the relative constancy of factor shares in the long run.

If firms were price takers, Fellner (1961) noted that some type of learning process had to be postulated to justify such firms investing in biased

We wish to acknowledge helpful comments from Jess Benhabib, William A. Barnett, Paul A. Samuelson, and Andrew Schotter. Address correspondence to: Ryuzo Sato, Director, The Center for Japan-U.S. Business and Economic Studies, Stern School of Business, 44 West Fourth Street, New York, NY 10012–1126, USA; e-mail: rsato@stern.nyu.edu.

technological innovations. Salter (1960) argued that firms should minimize total cost and not the cost of any input. Such cost minimization does not imply the Hicksian process of biased technological progress but Kennedy (1964) responded by postulating a static innovation-possibility frontier to reinstate the Hicksian hypothesis. Kennedy, however, took his theory to imply a rejection of the marginal productivity theory, but Samuelson (1965) showed that it is consistent with the neoclassical theory of production.

Nordhaus (1967) and Kamien and Schwartz (1969) generalized the model and made the position of the frontier dependent on the level of research expenditure. Although these papers provided a microeconomic foundation for investment in endogenous technological progress, they did not consider biased technological progress. An attempt to simultaneously determine the levels of and the bias of technological progress was made by Sato and Ramachandran (1987). They assumed that a monopolistic firm facing differential growth in input prices responds to the cost increases by increasing factor efficiency; using an optimal control model, they showed that both the rate and the bias of technological progress were such as to counterbalance the increases in input prices.

The second tradition can be traced back to the development of neoclassical theory of growth by Solow and Swan. Solow (1957), using the observable factor shares, estimated the rate of growth of total factor productivity and showed that more than half of the growth rate in per capital output should be attributed to technological change. In his empirical analysis, Solow had assumed Hicks neutral technological progress while his theoretical model has steady state if and only if technological progress is Harrod-neutral. Other forms of bias received less prominence in the voluminous literature on neoclassical growth models.

One of the reasons for this neglect is that there are some well-known problems in estimating biased technological progress. If the production function is Cobb-Douglas, then all factor-augmenting technological progress reduces to the Hicks-neutral form. If the production function has constant but nonunitary elasticity, then the Diamond-McFadden impossibility theorem states that the elasticity of substitution and bias cannot be estimated simultaneously. The theoretical explanation for the constancy of the factor shares seems to be beyond empirical verification.

To break this impasse, Sato (1970) derived the constant elasticity of derived demand (CEDD) production functions which had the convenient property that the elasticity of substitution was proportional to the factor share. It then was shown that the elasticities of derived demand for the factors are constant so that a simple regression analysis can be applied to estimate these elasticities. This property made it particularly useful for the estimation of the biased technological progress. The data from the U.S. nonfarm sector for 1909 to 1960 was used to test how well this function fitted relative to Cobb-Douglas and other forms. Rejection of the Cobb-Douglas form, which precluded any meaningful discussion of bias, set the stage for the estimation of biased technological progress using CEDD functions and comparing it with estimates using CES function. It is shown that a function with variable elasticity of substitution [$E_x(\alpha) = 1.558\alpha$, where x is the capital/labor ratio and α is the share of capital] had more explanatory power than the CES functions with reasonable values for the elasticity of substitution. Further, technological progress is shown to be biased to labor savings.

But the analysis by Sato (1970) was not embedded in a model of economic growth. We construct a model with endogenous technological progress that has two properties sought in growth models:

1. It shows the possibility of a steady state with biased technological progress, where the bias will depend on the parameters of the model. This constrasts with the earlier neoclassical growth models which have steady state only if technological progress is Harrod-neutral.
2. The motivation for introducing biased technological progress in growth models was to explain the constancy of factor share and, in this model, the factor shares are constant in the steady state.

2. Applying Hetrogeneous Capital Goods Model to Technological Progress

Samuelson and Solow (1956) extended the Ramsey model to the case of many capital goods. They assumed that there are n goods that can be used for production or consumption and that the stock of these goods determines the output through a general transformation relation:

$$C_1 + \dot{S}_1 = f(S_1 \dots S_n;\ C_2 + \dot{S}_2 \dots C_n + \dot{S}_n),$$

where S_i, $i = 1 \dots n$, are stocks of n goods and C_i is the rate of consumption of the goods in a period and $\dot{S}_i$ is the additions to the stock of the goods.

The utility $U(C_1 \dots C_n)$ is a concave function in the consumption of n goods and the problem is to maximize, using calculus of variations,

$$\int_0^{\infty} U(C_1 \dots C_n)\, dt \quad \text{subject to} \quad f(S_1 \dots S_n;\ C_2 + \dot{S}_2 \dots C_n + \dot{S}_n)$$
$$- C_1 - \dot{S}_1 = 0 \quad \text{and} \quad S_i\,(0) = \text{initial stock},\ i = 1 \dots n.$$

They assumed the Legendre condition that $((\partial^2 f)/(\partial C_i \partial C_j))$ is negative definite but did not explicitly discuss the Jacobi and Weierstrauss conditions [Gelfand and Fomin (1967, pp. 97–149)].

In applying their analysis to the problem of technological progress, we specialize it to the case of four variables. One represents the stock of a capital good, K. The second is labor that grows at a constant rate, n. The third and fourth variables are the effciency levels, A and B, of K and L, respectively. We now write the transformation function, using their suggestion (1967, p. 285) as

$$C_1 = \tilde{f}(K,\ L;\ A,\ B;\ \dot{K},\ \dot{L},\ \dot{A},\ \dot{B}),$$

and the problem is to maximize $\int_0^{\infty} U(C)\, dt$ subject to the transformation function and initial conditions.

As with the Samuelson-Solow model, various specializations are possible. We can rewrite the transformation function as $\tilde{F}[(AK), (BL), (A \cdot K), (B \cdot L)]$ and take $\tilde{F}$ to be homogenous of degree one in AK and BL or of two in A, K, B, and L. It can be shown that the Legendre condition is satisified for this problem where AK (BL) is treated as one variable but, because the increasing returns already noticed in the case where A, K, B, and L are considered as separate variables, it is not possible to establish the sufficiency conditions in the general case. We reformulate the problem in terms of optimal control with a specialized version of the transformation function and establish the concavity of the maximized Hamiltonian (known to be a weaker condition than those stated above) to establish the sufficiency condition.

3. Formulation of the Problem using Optimal Control

The transformation function is now written as

$$C_1 = F(AK, BL) - AK\phi_1\left(\frac{\dot{A}}{A}\right) - BL\phi_2\left(\frac{\dot{B}}{B}\right) - AK\phi_3\left(\frac{\dot{K}}{K}\right). \tag{1}$$

This brings the model in line with the production-function approach common in microeconomic models and neoclassical growth theory. $F(\cdot)$ is the production function of the numeraire commodity that can be either consumed or used for increasing the stock of the inputs other than labor.

The function ϕ_i, $i = 1$ or 2, can be thought of as inverses of the technological progress functions common in endogenous growth models [Sato and Ramachandran (1987)]; if we fix the total resources devoted to increasing A and B, then they will jointly define the Kennedy-Weizacker-Samuelson technological progress frontier.

If the expenditure on research intended to increase the efficiency of a factor is ϕ_i, $i = 1, 2$, then the rate of technological progress, $\dot{A}/A$ or $\dot{B}/B$, is taken as a function of ϕ_i. If current expenditures on research increase the efficiency of an input, and if it retains this efficiency level into the future, then an expenditure today leads to an incremental stream of income that extends to the planning horizon of the firm. The net discounted return from the investment may be so high that there is no interior solution to the allocation problem; the system moves to the corner with all of the output devoted to research as has been noticed in earlier models of endogenous technological progress. The economically uninteresting corner solution is avoided by assuming that the generation of technological progress has counterbalancing diminishing returns. This is partly achieved by assuming that the technological progress functions are concave; increases in the rate of expenditure bring about smaller increases in the rate of growth of efficiency, but, in models of optimal technological progress with infinite horizon, an additional assumption that strengthens the diminishing returns is necessary for the system to have an interior solution. Sato (1996) showed that the necessary condition for the existence of an interior steady state is that C_1 is (at least asymptotically) homogeneous of degree 2 with respect to its variables, A, K, B, L, $\dot{A}$, $\dot{K}$, $\dot{B}$, and $\dot{L}$.

We assume that the efficiency factor increases through investment in applied research and that it takes resources to transfer the new technology to the physical units such as labor and capital. As the quantity of an input in efficiency units increases, more resources are needed for generating incremental increases in efficiency; in other words, the expenditures, on entering the technological progress functions, are deflated by the quantity of inputs in efficiency units.

Let the numeraire good, $Y = F(AK, BL)$, allocated to increasing the efficiency of K and L be

$$M_i = m_i Y, \quad 0 < m_i < 1, \quad i = 1, 2. \tag{2}$$

Rates of growth of the efficiency of inputs are given by technological progress functions as linear homogeneous functions of the expenditure per unit of input and the level of technological progress of that factor:

$$\dot{A} = H_1\left[\frac{m_1 Y}{K}, A\right] = Ah_1\left[\frac{m_1 Y}{AK}\right]$$

and

$$\dot{B} = H_2\left[\frac{m_2 Y}{L}, B\right] = Bh_2\left[\frac{m_1 Y}{BL}\right].$$

The economics assumption behind these equations is that rate of growth of efficiency increases with expenditure per unit of input. Higher levels of efficiency would lead to a higher rate of growth but the proportionate growth rate of efficiency will decrease as the level of factor augmentation increases. These equations can be written as

$$\frac{\dot{A}}{A} = h_1\left(\frac{m_1 Y}{AK}\right) \tag{3a}$$

$$\frac{\dot{B}}{B} = h_2\left(\frac{m_2 Y}{BL}\right) \tag{3b}$$

where, because of the concavity assumption, $h_i'' < 0 < h_i'$, $i = 1, 2$.

Capital in efficiency terms, AK, increases either because of the increase in A or in K. Following Ramsey (1928), we assume that the saving rate is determined by the economy as a solution to the optimization problem. Under the assumption of one malleable output, most growth models of the Solow-Swan type assume that a unit of savings and investment will lead to a unit increase in capital; this assumes a linear transformation curve between consumption goods and capital goods.

Combining the traditional assumption with technological progress functions implies that an allocation of a unit of resources would bring about a unit increase in K whereas additional increases in research expenditures will bring about only smaller and smaller increases in A. These models have an inherent bias against investment in capital augmenting technological progress. It is uneconomic to enhance AK by increasing A, and the model leads to Harrod-neutral technological progress in equilibrium. Following Liviatan and Samuelson (1969), we assume that the transformation of consumption goods into capital goods is nonlinear, with the function taking a form similar to that of technological progress functions:

$$\dot{K} = H_3\left[\frac{M_3}{A}, K\right] = H_3\left[\frac{m_3 Y}{A}, K\right] = Kh_3\left[\frac{m_3 Y}{AK}\right]$$

or

$$\frac{\dot{K}}{K} = h_3\left[\frac{m_3 Y}{AK}\right]. \tag{4}$$

In this model, the increase in K and A are both subject to diminishing returns and the bias against capital- augmenting technological progress is eliminated. Notice that the forms of the technological progress functions are such that it satisfies the necessary condition stated by Sato (1996).

Labor grows at a constant proportionate rate,

$$\dot{L}/L = n. \tag{5}$$

Output can grow if A, K, B, or L increases. Increase in L is taken to be exogenous whereas increases in the other three require allocation of current output, and so, Y is allocated between consumption, savings, and expenditures on research:

$$Y = C + (m_1 + m_2 + m_3)Y. \tag{6}$$

We assume a linear utility function with per-capita consumption as the argument:

$$c(t) = \frac{C(t)}{L(t)} = \frac{Y(t)}{L(t)}(1 - m_1 - m_2 - m_3), \tag{7}$$

where t is the time and the second equality follows from (6).

The society chooses the three control variables, m_1, m_2, and m_3 so as to maximize the discounted value of the sum of utility over an infinite time. After substituting $L(t) = L_0 e^{nt} = e^{nt}$ in (7) (with L_0 equal to unity) and using (1), (3), and (4), the maximization problem can be written as

$$\underset{\{m_1, m_2, m_3\}}{\text{Maximize}} \int_0^\infty e^{-(\rho+n)t} F(AK, BL)(1 - m_1 - m_2 - m_3)\, dt \tag{8}$$

subject to

$$\frac{\dot{A}}{A} + \frac{\dot{K}}{K} = h_1\left[\frac{m_1 F(AK, BL)}{AK}\right] + h_3\left[\frac{m_3 F(AK, BL)}{AK}\right]$$

and

$$\frac{\dot{B}}{B} + \frac{\dot{L}}{L} = h_2\left[\frac{m_2 F(AK, BL)}{BL}\right] + n.$$

The current-value Hamiltonian can be written as

$$\tilde{H} = F(AK, BL)(1 - m_1 - m_2 - m_3) + P_1 AK[h_1() + h_3()] + P_2 BL[h_2() + n]. \quad (9)$$

Notice that we are taking AK and BL to be the two state variables and P_1 and P_2 to be the two costate variables. In analyzing this model, we first establish that a steady state with constant AK/BL is consistent with this model and then we use the steady-state growth rate as an auxiliary variable to transform the model into one that is more tractable for considering stability.

The instantaneous values of the control variables m_1, m_2, and m_3 are those that maximize the value of H; differentiating H partially with respect to these variables and the first-order conditions can be written as

$$P_1 h_1'[m_1 F(1, BL/AK)] = 1, \quad \textbf{(10a)}$$

$$P_2 h_2'[m_2 F(AK/BL, 1)] = 1, \quad \textbf{(10b)}$$

and

$$P_1 h_3'[m_3 F(1, BL/AK)] = 1, \quad \textbf{(10c)}$$

See Appendix A for details.

The values of m_i, $i = 1$, 2, and 3, that are obtained by solving (10) are written as $\tilde{m}_i$. Notice that $\tilde{m}_1$ and $\tilde{m}_3$ are functions P_1 and AK/BL only and m_2 of P_2 and AK/BL; this separability will be of use in drawing phase diagrams for the model. The equations of motion of the two costate variables, obtained from the partial differential of the Hamiltonian with respect to the corresponding state variables, are given by

$$\dot{P}_1 = -F_1 + P_1(\rho + n - h_1 - h_3) + (\tilde{m}_1 + \tilde{m}_3)(F/AK), \quad \textbf{(11a)}$$

$$\dot{P}_2 = -F_2 + P_2(\rho - h_2) + (\tilde{m}_2 F/BL). \quad \textbf{(11b)}$$

If AK/BL is a constant at the steady state, then

$$\left(\frac{\dot{A}}{A}\right)^* + \left(\frac{\dot{K}}{K}\right)^* = h_1^* + h_3^* = \left(\frac{\dot{B}}{B}\right)^* + \left(\frac{\dot{L}}{L}\right)^* = h_2^* + n = \left(\frac{\dot{Y}}{Y}\right)^*, \quad \textbf{(12)}$$

where * indicates the values at steady state. Substituting (12) into (11) and setting $\dot{P}_i = 0$, $i = 1, 2$, we get

$$P_1^*(\rho - h_2^*) = F_1 - (\tilde{m}_1^* + \tilde{m}_3^*)\frac{F}{AK} \quad \textbf{(13a)}$$

and

$$P_2^*(\rho - h_2^*) = F_2 - \tilde{m}_2^* \frac{F}{BL} \cdot \quad \textbf{(13b)}$$

Multiplying (13a) by AK and (13b) by BL, adding the two equations, and dividing both sides of the equality by Y and by $(\rho - h_2^*)$, we get

$$\frac{P_1^* \cdot AK + P_2^* \cdot BL}{Y} = \frac{1 - \tilde{m}_1^* - \tilde{m}_2^* - \tilde{m}_3^*}{\rho - h_2^*}. \quad \textbf{(13c)}$$

Because of (12), AK/Y and BL/Y are constants and so are the values of the control variables. Therefore, the equation is consistent with the assumption that AK/BL, P_1^*, and P_2^* are constants.

For an economic interpretation of (13c), note that the numerator on the righthand side is the steady-state consumption out of a unit of output, c^* for example. Then, per-capita consumption is $c^* Y/L$. Because Y increases at a rate $h_2^* + n$ while labor increases at a rate n, the per-capita income increases at a rate h_2^* and can be written as $c^* \cdot e^{h_2^* t}$. This consumption stream discounted from zero to infinity at a social discount rate of ρ has a present value $c^*/(\rho - h_2^*)$. The numerator on the right-hand side is the stocks of physical and human capital in efficiency units evaluated at their shadow prices. Hence (13c) states that the asset per unit of output equals the discounted value of consumption per person.

4. Analysis of Stability

To study stability, we introduce an auxiliary variable E whose rate of growth is equal to that of the steady-state growth rate of Y,

$$\dot{E}/E = \dot{Y}^*/Y^* = \varepsilon, \quad \textbf{(14)}$$

and redefine the state variables as

$$g_1 = AK/E, \quad g_2 = BL/E. \quad \textbf{(15)}$$

The production function and per-capita consumption can be written as functions of g_1 and g_2:

$$Y = EF\left(\frac{AK}{E}, \frac{BL}{E}\right) = EF(g_1, g_2)$$

and

$$c = \frac{C}{L} = \frac{E}{L} F(g_1, g_2)(1 - m_1 - m_2 - m_3).$$

Further, the technological progress functions can be expressed as

$$\frac{\dot{A}}{A} = h_1\left[\frac{m_1 F(g_1, g_2)}{g_1}\right], \quad \textbf{(16a)}$$

$$\frac{\dot{B}}{B} = h_2\left[\frac{m_2 F(g_1, g_2)}{g_2}\right], \quad \textbf{(16b)}$$

and

$$\frac{\dot{K}}{K} = h_3\left[\frac{m_3 F(g_1, g_2)}{g_1}\right]. \quad \textbf{(16c)}$$

The current-value Hamiltonian now can be written as

$$H = F(g_1, g_2)(1 - m_1 - m_2 - m_3) + P_1 g_1(h_1 + h_3 - \varepsilon) + P_2 g_2(h_2 + n - \varepsilon). \quad (17)$$

(See Appendix B.) The first-order conditions are given by equations similar in form to (10) with the appropriate transformation of variables. The equations of motion of the state and costate variables are given by

$$\dot{g}_1 = (h_1 + h_3 - \varepsilon) g_1, \quad (18a)$$

$$\dot{g}_2 = (h_2 + n - \varepsilon) g_2, \quad (18b)$$

$$\dot{P}_1 = P_1(\rho + n - h_1 - h_3) - F_{g1} + (\tilde{m}_1 + \tilde{m}_3)\frac{F}{g_1}, \quad (18c)$$

$$\dot{P}_2 = P_2(\rho - h_2) - F_{g2} + \tilde{m}_2 \frac{F}{g_2}. \quad (18d)$$

Except for a change of variables, the model is the same as the one in Section 2 and so, the properties of the model including that of the steady state are the same. The stability of the system can be examined by evaluating the Jacobian of the system of differential equations (18) at the steady state and calculating the characteristic roots. Direct calculation shows that one of the characteristic roots is zero, indicating that the dimensions of the system can be reduced once more.

We define a new variable $\theta = (g_1/g_2) = (AK/BL)$. The equations (18a–d) are replaced by

$$\frac{\dot{\theta}}{\theta} = h_1 \left[\frac{\tilde{m}_1 f(\theta)}{\theta}\right] + h_3 \left[\frac{\tilde{m}_3 f(\theta)}{\theta}\right] - h_2[\tilde{m}_2 f(\theta)] - n \quad (19a)$$

where

$$\frac{F(g_1, g_2)}{g_2} = F(g_1/g_2, 1) = F(\theta, 1) = f(\theta), \text{ and}$$

$$\dot{P}_1 = -f'(\theta) + P_1\{\rho + n - h_1 - h_3\} + (\tilde{m}_1 + \tilde{m}_3)\frac{f(\theta)}{\theta} \quad (19b)$$

$$\dot{P}_2 = -\{f(\theta) - \theta f'(\theta)\} + P_2\{\rho - h_2\} + \tilde{m}_2 f(\theta) \quad (19c)$$

The Jacobian of the system evaluated at the steady state is

$$\begin{bmatrix} 0 & v_1 & -v_2 \\ -f' & \rho - h_2^* & 0 \\ \theta^* f'' & 0 & \rho - h_2^* \end{bmatrix}$$

where

$$v_1 = -\frac{\theta^*}{P_1^*}\left[\frac{(h_1')^2}{h_1''} + \frac{(h_3')^2}{h_3'}\right]^* > 0$$

and

$$v_2 = -\frac{\theta^*}{P_2^*}\left[\frac{(h_3')^2}{h_2''}\right]^* > 0.$$

(See Appendix C for details of calculation.)

The characteristic equation can be written as

$$(\rho - h_2^* - \lambda)\{-\lambda(\rho - h_2^* - \lambda) + v_1 f'' + v_2 f''\theta\} = 0. \tag{20}$$

The three characteristic roots are

$$\lambda_1 = \rho - h_2^*,$$

$$\lambda_2 = \frac{\rho - h_2^* + \sqrt{(\rho - h_2^*)^2 - 4f''(v_1 + v_2\theta)}}{2},$$

$$\lambda_3 = \rho - h_2^* - \sqrt{\frac{(\rho - h_2^*)^2 - 4f''(v_1 + v_2\theta)}{2}}.$$

We assume that $\rho > h_2^*$; hence λ_1 and λ_2 are positive and λ_3 is negative. We have a saddle point of type 2.

The stability analysis can be illustrated using the phase diagram (Figure 1). Even though the diagram in its full generality is three-dimensional—corresponding to the variables θ, P_1, and P_2—it can be reduced to a series of two-dimensional cross sections.

First, as noted in the discussion of (10), $\tilde{m}_1$ and $\tilde{m}_3$ are functions of θ and P_1 only; hence $\dot{P}_1$ is a function of θ and P_1 only. Next, $\tilde{m}_2$ is a function of P_2 and θ; for given value of P_2, it is a function of θ. Hence, from equations (19), we see that for a given

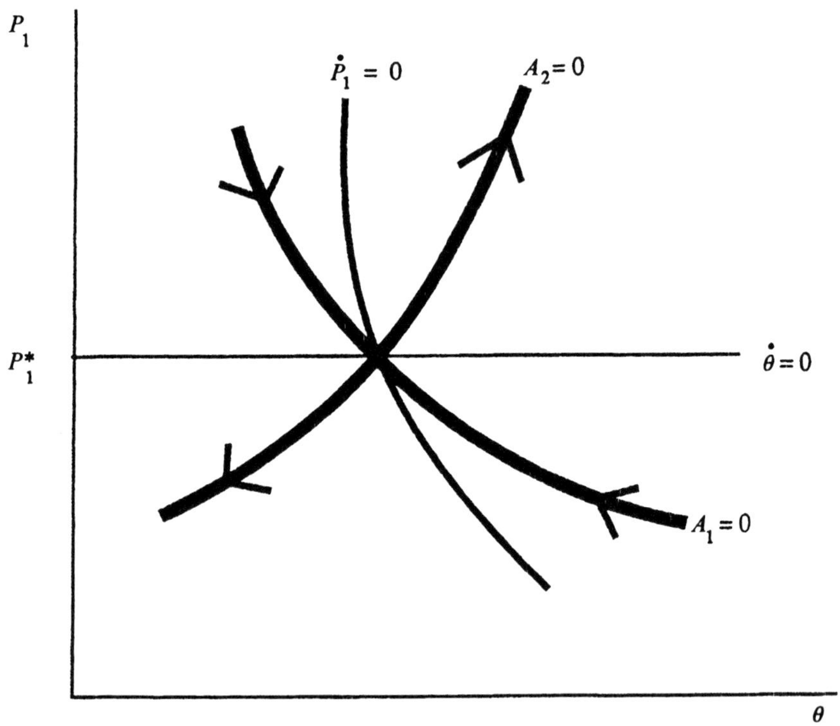

FIGURE 1. Phase diagram in $P_1 - \theta$ plane.

value of P_2, $\dot{\theta}$ are functions of P_1 and θ. Hence, we can fix the value of P_2 and study the motion of the system in the cross section of the phase diagram.[2]

The Jacobian of the reduced system is

$$\begin{bmatrix} 0 & v_1 \\ -f'' & \rho - \beta \end{bmatrix},$$

and the roots of the characteristic equation is given by

$$\lambda = \frac{(\rho - \beta) \pm \sqrt{(\rho - \beta)^2 - 4f''v_1}}{2}.$$

One of the roots is positive and the other is negative, showing saddle-point stability in the cross section. As stated in note 2, the value of θ corresponding to the singular points of these equations will vary from cross section to cross section. The value of P_2 corresponding to the steady state of the model is already derived. For that value of P_2, the system will show the convergence to the steady state in the cross-sectional diagram.

5. Conclusion

One of the paradoxes of the neoclassical model of growth is that it has a steady state only if technological progress is Harrod-neutral; a positive rate of labor augmentation has the same net consequence as a higher rate of growth of labor except that the per-capita output in steady state increases at the rate of increase in labor efficiency.

In contrast, this model has an interior solution and a steady state in which there is positive accumulation of capital and increases in the efficiency of both capital and labor. A comparison of the two models seems to indicate that the crucial difference is in the assumption about the effect of expenditure on investment and in increasing the efficiency of capital. Models of optimal endogenous technological progress assume that the expenditure on research has diminishing returns; without this assumption the models will explode. However, investment is assumed to increase capital in a linear manner. Under these assumptions, it is uneconomic for the system to increase AK by increasing A instead of K. The bias of the model toward Harrod-neutral technological progress is obvious.

We followed Liviatan and Samuelson (1969) in assuming a nonlinear transformation of the output to capital good. The output is taken as malleable, capable of being used both as a consumption good and a capital good but there is an implicit transformation process in converting the output into capital good and this transformation function is concave and nonlinear (as in the traditional neoclassical models). Further, we assumed, as in other models of endogeneous technological progress, that the technical progress functions are concave.

At steady state, the ratio of inputs measured in efficiency terms will remain constant and with it the factor shares. If one assumes a CEDD production function, then the elasticity of subsitution is also a constant at steady state. However, as the system converges to the steady state, both factor share and the elasticity of substitution will vary. Thus, it provides a growth theoretical model that is consistent with the empirical work of Sato (1970).

Notes

1. Ironically, the most recent concern is to explain the observed fall in the wages of relatively unskilled workers in the industrial nations. As a survey of the literature shows, some attribute it to the working of the factor price equalization theorem whereas others attribute it to shifts in technology [Burtless (1995)].
2. Note that θ is a function of $\tilde{m}(\theta, P_2)$ so that the singular point of the reduced system is dependent on P_2. As we move over the cross sections, the $\dot{P}_1 = 0$ curve does not shift but the $\dot{\theta} = 0$ curve will. Similarly, we can take cross sections of the phase diagram, keeping P_1 constant, and draw the $\dot{\theta} = 0$ and $\dot{P}_2 = 0$ curves.

References

Burtless, G. (1995) International trade and the rise in earnings inequality. *Journal of Economic Literature* 32, 800–816.

Fellner, W. (1961) Two propositions in the theory of induced innovations. *Economic Journal* 71, 305–308.

Gelfand, I.M. & S.V. Fomin (1967) *Calculus of Variations*. Englewood Cliffs, NJ: Prentice-Hall.

Hicks, J.R. (1963) *The Theory of Wages*. London: Macmillan.

Kamien, M.I. & N.L. Schwartz (1969) Induced factor augmenting technical progress from a microeconomic viewpoint. *Econometrica* 37, 668–684.

Kennedy, C. (1964) Induced bias in innovation and the theory of distribution. *Economic Journal* 37, 541–547.

Liviatan, N. & P.A. Samuelson (1969) Notes on turnpikes: Stable and unstable. *Journal of Economic Theory* 1, 454–475.

Nordhouse, W.D. (1967) The optimal rate and direction of technical change. In K. Shell (ed.), *Essays in the Theory of Optimal Economic Growth*, pp. 53–66. Cambridge, MA: MIT Press.

Ramsey, F.P. (1928) A mathematical theory of saving. *Economic Journal* 38, 543–559.

Salter, W.E.G. (1960) *Productivity and Technical Change*. Cambridge, England: Cambridge University Press.

Samuelson, P.A. (1965) A theory of induced innovations along Kennedy-Wizaker lines. *Review of Economics and Statistics* 47, 343–356.

Samuelson, P.A. & R.M. Solow (1956) A complete capital model involving heterogenous capital goods. *Quarterly Journal of Economics* 70, 537–562.

Sato, R. (1970) The estimation of biased technical progress and the production function. *International Economic Review* 11, 179–208.

Sato, R (1996) A note on modeling endogenous growth. *Keio Economic Studies* 39, 343–356.

Sato, R. & R. Ramachandran (1987) Factor price variation and the Hicksian hypothesis: A microeconomic model, *Oxford Economic Papers* 39, 343–356.

Solow, R. (1957) Technical change and the aggregate production function. *Review of Economics and Statistics* 39, 312–320.

Appendix A: Derivation of Basic Model

From equation (8), the Hamiltonian of the system can be written as

$$\tilde{H} = e^{-(\rho+n)t}F(AK, BL)(1 - m_1 - m_2 - m_3) + q_2BL \times \left\{h_2\left[\frac{m_2F(AK, BL)}{BL}\right] + n\right\} + q_1AK \times \left\{h_1\left[\frac{m_1F(AK, BL)}{AK}\right] + h_3\left[\frac{m_3F(AK, BL)}{AK}\right]\right\}. \quad \text{(A.1)}$$

The first-order conditions are as follows:

$$\tilde{H}_{m_1} = 0 \Rightarrow q_1 e^{(\rho+n)t} h_1' = 1, \quad \text{(A.2a)}$$

$$\tilde{H}_{m_2} = 0 \Rightarrow q_2 e^{(\rho+n)t} h_2' = 1, \quad \text{(A.2b)}$$

$$\tilde{H}_{m_3} = 0 \Rightarrow q_1 e^{(\rho+n)t} h_3' = 1, \quad \text{(A.2c)}$$

$$\dot{q}_1 = -\frac{\partial \tilde{H}}{\partial (AK)} = -e^{-(\rho+n)t} F_1 (1 - m_1 - m_2 - m_3)$$

$$-q_1 (h_1' m_1 + h_3' m_3) \frac{F_1 AK - F}{AK} - q_1 (h_1 + h_3) - q_2 h_2' m_2 F_1, \quad \text{(A.2d)}$$

$$\dot{q}_2 = -\frac{\partial \tilde{H}}{\partial (BL)} = -e^{-(\rho+n)t} F_2 (1 - m_1 - m_2 - m_3)$$

$$-q_1 (h_1' m_1 + h_3' m_3) F_2 - q_2 (h_2 + n) - q_2 h_2' m_2 \frac{F_2 BL - F}{BL}. \quad \text{(A.2e)}$$

Let

$$P_1 = q_1 e^{(\rho+n)t}, \quad \text{(A.3a)}$$

$$P_2 = q_2 e^{(\rho+n)t}; \quad \text{(A.3b)}$$

then,

$$\dot{P}_1 = \dot{q}_1 e^{(\rho+n)t} + (\rho + n) P_1, \quad \text{(A.4a)}$$

$$\dot{P}_2 = \dot{q}_2 e^{(\rho+n)t} + (\rho + n) P_2. \quad \text{(A.4b)}$$

Using (A.3), from (A.2a), (A.2b), and (A.2c), we get

$$P_1 h_1' \left[\frac{m_1 F(AK, BL)}{AK} \right] = 1, \quad \text{(A.5a)}$$

$$P_2 h_2' \left[\frac{m_2 F(AK, BL)}{BL} \right] = 1, \quad \text{(A.5b)}$$

$$P_1 h_3' \left[\frac{m_3 F(AK, BL)}{BL} \right] = 1. \quad \text{(A.5c)}$$

Multiplying (A.2d) and (A.2e) by $e^{(\rho+n)t}$ and substituting the two equations into (A.4a) and (A.4b), respectively, we get

$$\dot{P}_1 = -F_1 + P_1 \left\{ \rho + n + h_1 \left[\frac{\tilde{m}_1 F(AK, BL)}{AK} \right] - h_3 \left[\frac{\tilde{m}_3 F(AK, BL)}{AK} \right] \right\}$$

$$+ (\tilde{m}_1 + \tilde{m}_3) \frac{F(AK, BL)}{AK} \quad \text{(A.5d)}$$

$$\dot{P}_2 = -F_2 + P_2\left\{\rho - h_2\left[\frac{\tilde{m}_2 F(AK, BL)}{BL}\right]\right\} + \tilde{m}_2 \frac{F(AK, BL)}{BL}. \tag{A.5e}$$

Appendix B: Model with Auxiliary Variable

After the transformations [see equations (14)–(16)], the original problem [equation (8)] can be written as

$$\max_{\{m_1, m_2, m_3\}} \int_0^\infty e^{-(\rho+n-\varepsilon)t} F(g_1, g_2)(1 - m_1 - m_2 - m_3)dt$$

subject to

$$\frac{\dot{g}_1}{g_1} = h_1\left[\frac{m_1 F(g_1, g_2)}{g_1}\right] + h_3\left[\frac{m_3 F(g_1, g_2)}{g_1}\right] - \varepsilon, \tag{B.1a}$$

$$\frac{\dot{g}_2}{g_2} = h_2\left[\frac{m_2 F(g_1, g_2)}{g_2}\right] + n - \varepsilon. \tag{B.1b}$$

The current-value Hamiltonian is

$$H = F(g_1, g_2)(1 - m_1 - m_2 - m_3) + P_2 g_2\left\{h_2\left[\frac{m_2 F(g_1, g_2)}{g_2}\right] + n - \varepsilon\right\}$$
$$+ P_1 g_1\left\{h_1\left[\frac{m_1 F(g_1, g_2)}{g_1}\right] + h_3\left[\frac{m_3 F(g_1, g_2)}{g_1}\right] - \varepsilon\right\}. \tag{B.2}$$

The first-order conditions are

$$P_1 h_1'\left[\frac{m_1 F(g_1, g_2)}{g_1}\right] = 1, \tag{B.3a}$$

$$P_2 h_2'\left[\frac{m_2 F(g_1, g_2)}{g_2}\right] = 1, \tag{B.3b}$$

$$P_1 h_3'\left[\frac{m_3 F(g_1, g_2)}{g_1}\right] = 1\cdot \tag{B.3c}$$

The equation of motion of the costate variables can be written as

$$\dot{P}_1 = P_1(\rho + n - \varepsilon) - \frac{\partial H}{\partial g_1}$$
$$= P_1\left\{\rho + n - h_1\left[\frac{\tilde{m}_1 F(g_1, g_2)}{g_1}\right] - h_3\left[\frac{\tilde{m}_3 F(g_1, g_2)}{g_1}\right]\right\}$$
$$- F_{g1} + (\tilde{m}_1 + \tilde{m}_3)\frac{F(g_1, g_2)}{g_1}, \tag{B.3d}$$

$$\dot{P}_2 = P_2(\rho + n - \varepsilon) - \frac{\partial H}{\partial g_2}$$
$$= P_2\left\{\rho - h_2\left[\frac{\tilde{m}_2 F(g_1, g_2)}{g_2}\right]\right\} - F_{g2} + \tilde{m}_2 \frac{F(g_1, g_2)}{g_2}. \tag{B.3e}$$

Appendix C: Stability

Define $\theta = (g_1/g_2) = (AK/BL)$ and remember that

$$\frac{F(AK, BL)}{AK} = \frac{F(g_1, g_2)}{g_1} = \frac{F(\theta, 1)}{\theta} = \frac{f(\theta)}{\theta},$$
$$\frac{F(AK, BL)}{BL} = \frac{F(g_1, g_2)}{g_2} = F(\theta, 1) = f(\theta),$$
$$F_1 = F_{g_1} = f'(\theta),$$

and

$$F_2 = F_{g_2} = f(\theta) - \theta f'(\theta).$$

Then, from either (A.5) or (B.3), we can get

$$P_1 h_1' \left[\frac{m_1 f(\theta)}{\theta}\right] = 1, \tag{C.1a}$$

$$P_2 h_2'[m_2 f(\theta)] = 1, \tag{C.1b}$$

$$P_1 h_3' \left[\frac{m_3 f(\theta)}{\theta}\right] = 1. \tag{C.1c}$$

Solving (C.1a), (C.1b), and (C.1c), we get $\tilde{m}_i (i = 1, 2, 3)$ and substitute into the equations of motion for costate variables:

$$\dot{P}_1 = -f'(\theta) + P_1\left\{\rho + n - h_1\left[\frac{\tilde{m}_1 f(\theta)}{\theta}\right] - h_3\left[\frac{\tilde{m}_3 f(\theta)}{\theta}\right]\right\}$$
$$+(\tilde{m}_1 + \tilde{m}_3)\frac{f(\theta)}{\theta}, \tag{C.1d}$$

$$\dot{P}_2 = -[f(\theta) - \theta f'(\theta)] + P_2\{\rho - h_2[\tilde{m}_2 f(\theta)]\} + \tilde{m}_2 f(\theta). \tag{C.1e}$$

From the definition of $\theta = (AK/BL)$, we can get

$$\frac{\dot{\theta}}{\theta} = h_1\left[\frac{\tilde{m}_1 f(\theta)}{\theta}\right] + h_3\left[\frac{\tilde{m}_3 f(\theta)}{\theta}\right] - h_2[\tilde{m}_2 f(\theta)] - n. \tag{C.1f}$$

The stability conditions of the system in the neighborhood of the steady state can be determined by the characteristic roots of the Jacobian matrix,

$$J = \frac{D(\dot{\theta}, \dot{P}_1, \dot{P}_2)}{D(\theta, P_1, P_2)} = \begin{bmatrix} \frac{\partial \dot{\theta}}{\partial \theta} & \frac{\partial \dot{\theta}}{\partial P_1} & \frac{\partial \dot{\theta}}{\partial P_2} \\ \frac{\partial \dot{P}_1}{\partial \theta} & \frac{\partial \dot{P}_1}{\partial P_1} & \frac{\partial \dot{P}_1}{\partial P_2} \\ \frac{\partial \dot{P}_2}{\partial \theta} & \frac{\partial \dot{P}_2}{\partial P_1} & \frac{\partial \dot{P}_2}{\partial P_2} \end{bmatrix}. \tag{C.2}$$

Using equations (C.1a), (C.1b), and (C.1c), we get

$$\frac{\partial \tilde{m}_1}{\partial \theta} = -\frac{P_1 h_1'' \tilde{m}_1 (\theta f' - f)/\theta^2}{P_1 h_1'' f/\theta} = -\frac{\tilde{m}_1}{\theta f}(\theta f' - f), \tag{C.3a}$$

$$\frac{\partial \tilde{m}_1}{\partial P_1} = -h'/(P_1 h_1'' f/\theta) = -\theta h_1'/(P_1 h_1'' f), \tag{C.3b}$$

$$\frac{\partial \tilde{m}_1}{\partial P_2} = 0, \tag{C.3c}$$

$$\frac{\partial \tilde{m}_2}{\partial \theta} = -\frac{P_2 h_2'' \tilde{m}_2 f}{P_2 h_2'' f} = -\frac{\tilde{m}_2 f'}{f}, \tag{C.3d}$$

$$\frac{\partial \tilde{m}_2}{\partial P_1} = 0, \tag{C.3e}$$

$$\frac{\partial \tilde{m}_2}{\partial P_2} = -h_2'/P_2 h_2'' f, \tag{C.3f}$$

$$\frac{\partial m_3}{\partial \theta} = -\frac{P_1 h_3'' \tilde{m}_3 (0 f' - f)/\theta^2}{P_1 h_3'' f/\theta} = -\frac{\tilde{m}_3}{\theta f}(\theta f' - f), \tag{C.3g}$$

$$\frac{\partial \tilde{m}_3}{\partial P_1} = -h_3'/(P_1 h_3'' f/\theta) = -\theta h_3'/(P_1 h_3'' f), \tag{C.3h}$$

$$\frac{\partial \tilde{m}_3}{\partial P_2} = 0. \tag{C.3i}$$

Then,

$$\frac{\partial(\tilde{m}_1 f/\theta)}{\partial \theta} = \frac{\partial \tilde{m}_1}{\partial \theta}\frac{f}{\theta} + \tilde{m}_1(\theta f' - f)/\theta^2 = 0, \tag{C.4a}$$

$$\frac{\partial(\tilde{m}_1 f/\theta)}{\partial P_1} = \frac{\partial \tilde{m}_1}{\partial P_1}\frac{f}{\theta} = -h_1'/P_1 h_1'', \tag{C.4b}$$

$$\frac{\partial(\tilde{m}_1 f/\theta)}{\partial P_2} = 0, \tag{C.4c}$$

$$\frac{\partial(\tilde{m}_2 f)}{\partial \theta} = \frac{\partial \tilde{m}_2}{\partial \theta} f + \tilde{m}_2 f' = -\tilde{m}_2 f' + \tilde{m}_2 f' = 0, \tag{C.4d}$$

$$\frac{\partial(\tilde{m}_2 f)}{\partial P_1} = 0, \tag{C.4e}$$

$$\frac{\partial(\tilde{m}_2 f)}{\partial P_2} = \frac{\partial \tilde{m}_2}{\partial P_2} f = -h_2'/P_2 h_2'', \tag{C.4f}$$

$$\frac{\partial(\tilde{m}_3 f/\theta)}{\partial \theta} = \frac{\partial \tilde{m}_3}{\partial \theta}\frac{f}{\theta} + \tilde{m}_3(\theta f' - f)/\theta^2 = 0, \tag{C.4g}$$

$$\frac{\partial(\tilde{m}_3 f/\theta)}{\partial P_1} = \frac{\partial \tilde{m}_3}{\partial P_1}\frac{f}{\theta} = -h_3'/P_1 h_3'', \tag{C.4h}$$

$$\frac{\partial(\tilde{m}_3 f/\theta)}{\partial P_2} = 0, \tag{C.4i}$$

The elements of the Jacobian matrix, which are evaluated at the steady state, are as follows:

$$\left(\frac{\partial \dot{\theta}}{\partial \theta}\right)^* = \theta^* \left[h_1' \frac{\partial(\tilde{m}_1 f/\theta)}{\partial \theta} + h_3' \frac{\partial(\tilde{m}_3 f/\theta)}{\partial \theta} - h_2' \frac{\partial(\tilde{m}_2 f)}{\partial \theta}\right]^* = 0, \tag{C.5a}$$

$$\begin{aligned}\left(\frac{\partial \dot{\theta}}{\partial P_1}\right)^* &= \theta^* \left[h_1' \frac{\partial(\tilde{m}_1 f/\theta)}{\partial P_1} + h_3' \frac{\partial(\tilde{m}_3 f/\theta)}{\partial P_1} - h_2' \frac{\partial(\tilde{m}_2 f)}{\partial P_1}\right]^* \\ &= -\frac{\theta^*}{P_1^*}\left[\frac{(h_1')^2}{h_1''} + \frac{(h_3')^2}{h_3''}\right]^* = v_1,\end{aligned} \tag{C.5b}$$

$$\begin{aligned}\left(\frac{\partial \dot{\theta}}{\partial P_2}\right)^* &= \theta^* \left[h_1' \frac{\partial(\tilde{m}_1 f/\theta)}{\partial P_2} + h_3' \frac{\partial(\tilde{m}_3 f/\theta)}{\partial P_2} - h_2' \frac{\partial(\tilde{m}_2 f)}{\partial P_2}\right]^* \\ &= -\frac{\theta^*}{P_2^*}\left[\frac{(h_2')^2}{h_2''}\right]^* = -v_2,\end{aligned} \tag{C.5c}$$

$$\begin{aligned}\left(\frac{\partial \dot{P}_1}{\partial \theta}\right)^* &= -f'' + P_1^*\left[-h_1' \frac{\partial(\tilde{m}_1 f/\theta)}{\partial \theta} - h_3' \frac{\partial(\tilde{m}_3 f/\theta)}{\partial \theta}\right]^* + \frac{\partial(\tilde{m}_1 f/\theta)}{\partial \theta} \\ &\quad + \frac{\partial(\tilde{m}_3 f/\theta)}{\partial \theta} = -f''\end{aligned} \tag{C.5d}$$

(since $P_1 h_i' - 1 = 0$, $i = 1 \,\&\, 3$),

$$\begin{aligned}\left(\frac{\partial \dot{P}_1}{\partial P_1}\right)^* &= (\rho + n - h_1^* - h_3^*) - P_1^*\left[h_1' \frac{\partial(\tilde{m}_1 f/\theta)}{\partial P_1} + h_3' \frac{\partial(\tilde{m}_3 f/\theta)}{\partial P_1}\right]^* \\ &\quad + \frac{\partial(\tilde{m}_1 f/\theta)}{\partial P_1} + \frac{\partial(\tilde{m}_3 f/\theta)}{\partial P_1} = \rho - h_2^*\end{aligned} \tag{C.5e}$$

(at the steady state, $h_1^* + h_3^* = h_2^* + n$),

$$\left(\frac{\partial \dot{P}_1}{\partial P_2}\right)^* = -P_1^* \left[h_1' \frac{\partial(\tilde{m}_1 f/\theta)}{\partial P_2} + h_3' \frac{\partial(\tilde{m}_3 f/\theta)}{\partial P_2}\right]^* + \frac{\partial(\tilde{m}_1 f/\theta)}{\partial P_2} + \frac{\partial(\tilde{m}_3 f/\theta)}{\partial P_2} = 0, \tag{C.5f}$$

$$\left(\frac{\partial \dot{P}_2}{\partial \theta}\right)^* = \theta^* f'' - [(P_2 h_2')^* - 1]\frac{\partial(\tilde{m}_2 f)}{\partial \theta} = \theta^* f'', \tag{C.5g}$$

$$\left(\frac{\partial \dot{P}_2}{\partial P_1}\right)^* = -[(P_2 h_2')^* - 1]\frac{\partial(\tilde{m}_2 f)}{\partial P_1} = 0, \tag{C.5h}$$

$$\left(\frac{\partial \dot{P}_2}{\partial P_2}\right)^* = \rho - h_2^* - [(P_2 h_2')^* - 1]\frac{\partial(\tilde{m}_2 f)}{\partial P_2} = \rho - h_2^*. \tag{C.5i}$$

So, substituting (C.5) into (C.2), we get

$$J^* = \frac{D(\dot{\theta}, \dot{P}_1, \dot{P}_2)^*}{D(\theta, P_1, P_2)^*} = \begin{pmatrix} \frac{\partial \dot{\theta}}{\partial \theta} & \frac{\partial \dot{\theta}}{\partial P_1} & \frac{\partial \dot{\theta}}{\partial P_2} \\ \frac{\partial \dot{P}_1}{\partial \theta} & \frac{\partial \dot{P}_1}{\partial P_1} & \frac{\partial \dot{P}_1}{\partial P_2} \\ \frac{\partial \dot{P}_2}{\partial \theta} & \frac{\partial \dot{P}_2}{\partial P_1} & \frac{\partial \dot{P}_2}{\partial P_2} \end{pmatrix}^* = \begin{pmatrix} 0 & v_1 & -v_2 \\ -f'' & \rho - h_2^* & 0 \\ \theta^* f'' & 0 & \rho - h_2^* \end{pmatrix} \tag{C.2'}$$

where

$$v_1 = -\frac{\theta^*}{P_1^*}\left[\frac{(h_1')^2}{h_1''} + \frac{(h_3')^2}{h_3''}\right]^* > 0,$$

and

$$v_2 = -\frac{\theta^*}{P_2^*}\left[\frac{(h_2')^2}{h_2''}\right]^* > 0.$$

Chapter 4

A THREE SECTOR MODEL OF ENDOGENOUS HICKSIAN BIAS

1. Introduction

The theory of technical progress seems to follow two separate but independent traditions. In the earlier microeconomic tradition that can be traced to Hicks's *Theory of Wages* (1932), the bias is determined by the relative growth in input prices. This proposition was challenged by Salter (1960), but Kennedy (1964) responded by postulating a static innovation possibility frontier to explain factor shares, while Samuelson (1965) and Drandakis and Phelps (1966) incorporated the frontier within the marginal productivity analysis.

Nordhaus (1967) and Kamien and Schwartz (1969) developed microeconomic models in which the position of the innovation possibility frontier is endogenously determined. But their models did not incorporate an induced bias from differential growth in factor prices; this step was taken in Sato and Ramachandran (1987).

The other tradition can be attributed to Solow's (1956, 1957) papers, which laid the foundation for the neoclassical growth theory. Solow (1956) assumed a stationary linear homogeneous production function and showed that at low capital–labour ratios the savings will exceed what is needed to maintain the ratio in spite of depreciation and growth in the labour force. This will lead to capital-deepening and to increases in the per capita income until the steady state is reached; thereafter the ratios will remain constant. In contrast, Solow (1957) attributed most of the observed growth in per capital income to shifts in the production function.

At a theoretical level, it was argued that technical progress is consistent with the steady state of a neoclassical growth model only if technical progress is Harrod-neutral; even the new theories of endogenous growth assume Harrod neutrality.[1] But there is increasing recognition that the economy takes a long time to converge to a steady state and that many economies under observation may be going through such an adjustment, a point made in Sato (1963).

[1] Grossman and Helpman (1993) and Lucus (1993) provide surveys of the endogeneous growth models.

In this paper we develop a model of biased growth where, even in the steady state, the efficiencies of capital and labour are both increasing as a result of allocations of resources in the research sector.

2. Modelling Bias in Technological Change

Any theory of biased technical progress must accord with the stylized facts and meet the theoretical challenge of building a model using a set of acceptable assumptions.

Economic historians argue that in industrialized countries the rate of interest is oscillatory but the wage rate is steadily increasing. In the neoclassical model, if the economy is still in the process of converging to a steady state, then the wage rate will increase but the interest rate will decrease. Harrod-neutral technical progress, by increasing the efficiency of labour in the steady state, can explain both increasing wages and a constant interest rate, but it requires the heroic assumption that the economy has been in dynamic equilibrium during the last one hundred years.

However, before introducing bias, the Diamond–McFadden impossibility theorem, which states that elasticity of substitution and bias cannot be estimated simultaneously, has to be addressed. Sato (1970) used a constant elasticity of derived demand (CEDD) production function, whose elasticity of substitution is proportional to the factor share, to get round the estimation problem. This function has the convenient empirical property that the elasticities of derived demand are constant and can be estimated by a simple regression analysis. The data from the US non-farm sector for 1909–60 were used to fit the CEDD production function and the fit was compared with those of Cobb–Douglas and other functional forms. Rejection of the Cobb–Douglas (which precludes any meaningful discussion of bias) set the stage for an estimation of bias using the CEDD function. It was found that the CEDD function for which $E_X(\alpha) = 1.558$ (where x is the capital–labour ratio and α is the share of capital) has more explanatory power than the CES function with reasonable elasticities of substitution. Table 1 gives a summary of the estimation using the CEDD function. This estimation shows that the effect on interest rates of the growth in the capital–labour ratio is moderated by the bias in technical progress. As a purely empirical paper, however, Sato (1970) did not examine how technical progress could be generated through an optimal allocation of resources for investment in physical capital and research. A number of theoretical problems had to be solved before formulating such a model.

There is more than one possible explanation for the observed increases in productivity. Some recent papers attribute it to external economies created by the spillovers from investment in physical or human capital. Others attribute it to capital accumulation that continues indefinitely because of the properties of the production function. Even those who emphasize the role of R&D attribute growth in productivity to labour-augmenting innovations.

The conventional wisdom is that Harrod-neutral technical progress alone is consistent with the steady state in a neoclassical growth model. Sato (1996)

TABLE 1

	1910	1960	Growth rate
Capital	145,520	412,304	
Labour	68,831	111,881	
Capital–labour ratio, K/L	2.079	3.685	0.0115
Relative efficiency of capital, A/B	1 (by definition)	0.465	−0.0153
AK/BL	2.079	1.714	−0.0038
Average of $\dot{A}/A$ for the period			0.0074
Average of $\dot{B}/B$ for the period			0.0227

Source: Sato (1970, pp. 206–207).

sought to examine the sufficient conditions for a steady state when technical progress is factor-augmenting. The production function was formulated as

$$Y = F(AK, BL). \tag{1}$$

The "investment" function, $G(\dot{K}, \dot{L}, \dot{A}, \dot{B}; K, L, A, B)$, determines the allocation of homogeneous output not only to increase the capital in physical units, but also to increase the efficiencies of both inputs in efficiency units. (In the function $\dot{x}$ is the growth rate of variable x, $x = K, L, A, B$.) Consumption is then output less investment:

$$C = F(AK, BL) - G(\dot{K}, \dot{L}, \dot{A}, \dot{B}; K, L, A, B). \tag{2}$$

Given per capita, output $c = C/L$, the objective is to maximize the discounted value of the per capita output:

$$\max \int_0^{\infty} e^{-\rho t} c dt. \tag{3}$$

If $F(\ )$ is linear homogeneous in inputs measured in efficiency units, then it is homogeneous of degree 2 in A, K, B and L. If the model is to have a steady state in the sense that $\lim_{t\to\infty} (AK/BL)$ is a constant, then the *sufficient condition* is that $C = F(\ ) - G(\ )$ has to be homogeneous of degree 2 in the steady state. This requires that $G(\ )$ is asymptotically homogeneous of degree 2 with respect to its variables. Harrod-neutral technical progress is not the only case in which a steady state can exist.

Sato *et al.* (1999) sought to develop a growth model using the sufficiency condition formally stated in Sato (1996). Investment is made to increase both the physical quantity of capital and the efficiency of inputs. The rate of growth in efficiency of an input is a function of the investment to increase its efficiency and the expenditure on research; this was expressed, without loss of generality, as a fraction of the output, $m_i F$. But two restrictions have to be imposed on the technical progress functions, h_i, if the optimal control model is to have a steady state. First, the functions must be concave, or $h_i'' < 0 < h_i'$. Further, in models of optimal technical progress with infinite horizon, an additional assumption that strengthens the diminishing returns is necessary for the system to have an interior solution. As output increases as a result of the growth of inputs in efficiency units, fixed values of m_i will lead to even higher levels of expenditure on research.

This in turn will accelerate the growth of productivity. The explosive effect of this process can be contained if it is assumed that, as the stock of inputs in efficiency units increases, more resources are needed to generate further increases in efficiency. This is achieved in this paper by deflating m_iF by AK, the stock of capital measured in efficiency units.

In the paper we used a specific investment function, $G(\ )$, defined as

$$G(\dot{A}, \dot{K}, \dot{B}, \dot{L}, A, K, B, L) = AK\phi_1\left(\frac{\dot{A}}{A}\right) + BL\phi_2\left(\frac{\dot{B}}{B}\right) + AK\phi_3\left(\frac{\dot{K}}{K}\right),$$

where ϕ_i, $i = 1, 2, 3$ are homogeneous of degree 0. The analysis of the model showed that it has an interior solution in which the ratio of inputs measured in efficiency terms remain constant while all the inputs increase in physical and efficiency terms.

The obvious deficiency of this model is that it assumes that increases in A, K and L are achieved by the expenditure of one homogeneous output. In this paper we relax this assumption by assuming that there are three sectors and that the consumption good, produced with capital and labour, is used for consumption alone. A Kennedy–Samuelson–Weizacker type innovation possibility frontier, whose position is determined by the allocation of labour to the research sector, determines the rate of growth of productivity of each factor.

In the Solow–Swan models, the entire savings is invested in formation of physical capital; technical progress, if any, is exogenously given and has to be Harrod-neutral if a steady state is to exist. In such a state, the growth rate of capital can exceed that of labour as long as the difference equals the rate of labour-augmenting technical progress; if the growth rate of capital is 5% and the growth rate of labour 2%, then the needed growth rate in labour-augmenting technical progress is 3%.

If one believes in endogenous technical progress, one has to ask where the resources for generating technical progress are to be allocated. Should all the expenditure on research by firms in developing technologies be concentrated in improving labour productivity, or should it be used also to improve the productivity of other inputs? Returning to the numerical example above, would not the economy be better off if it could reduce the growth rate of capital to 4% and use the freed savings to increase the productivity of capital by 1.5% and that of labour by 3.5%? This is an empirical question that needs investigation. If one considers the economic development of Japan, it is clear that the country did benefit from the transfer of technology from the Western countries, particularly the United States; but it has also invested heavily, in recent years, in developing new technology in electronics and robotics.

The argument in the Sato *et al.* (1999) paper is that in neoclassical growth models there is an inherent bias against increasing the productivity of capital at the cost of its physical growth. These models assume that a unit of savings of the homogeneous output will always lead to an increase in one unit of physical capital, but additional allocations to research have diminishing returns because of the concavity of the technical progress functions. To remove this bias, the paper assumed a Leviathan–Samuelson type of nonlinear investment function and established the existence of a steady state with bias. In this paper we retain the traditional assumption of a linear investment function, but relax the assumption of a malleable output used for consumption and investment.

3. A Model of Biased Technical Progress

We assume that the economy is divided into three sectors. The first sector produces a consumption good, the second sector produces a capital good, and the third sector determines the position of the innovation possibility frontier. We assume that the total quantity of labour is given at any moment of time and that it is allocated among the three sectors:

$$L_C + L_K + L_R = L = L_0 e^{nt},$$

or, by dividing by L,

$$l_C + l_K + l_R = 1. \tag{4}$$

The consumption good is produced using capital and labour. The constant-returns-to-scale production function is

$$C = F(AK, BL_C), \tag{5}$$

where A and B are the efficiency factors of the inputs.

Capital good is produced using only labour:[2]

$$\frac{\dot{K}}{K} = \left(\frac{DL_K}{K}\right) - \delta = \left(\frac{Dl_K}{k}\right) - \delta, \tag{6}$$

where δ is the rate of depreciation, D is the efficiency of labour in the capital-goods-producing sector, and $k = K/L$.

The rate of factor augmentation is determined in the research sector according to a concave innovation possibility frontier,[3]

$$G_1(A, \dot{A}, B, \dot{B}, D, \dot{D}, E, \dot{E}, K, L_R) = 0,$$

where E is the efficiency of labour in the research sector and, as labour is the only input into the research sector, changes in E can be thought of as Harrod neutral technical progress.

In examining the properties of the model, we will begin by assuming that the production possibilities frontier is defined by the following function:[4]

[2] We are grateful to Paul Samuelson for pointing out the similarities between our formulation and his elaboration of Thünen's theory of capital in Samuelson (1983).

[3] The investment function $G_1()$ is similar to the investment function $G(\ )$ in equation (2) except that, because this is a three-sector model, it is the labour employed in the research sector, L_R, and not L, that is the relevant variable.

[4] We are thankful to an anonymous referee for pointing out that the frontier reduces to the Uzawa-Lucus formulation if we set h_B, h_D, h_E and h_0 equal to zero. The referee also pointed out that the central message of the paper is not affected by the simplification. We recalculated the model under these assumptions; the Lagrangian given in (9) will be simplified as one state variable, g_4, is eliminated. Most of the results relating to the other state variables are not affected, as the referee correctly pointed out. While that provides evidence for the robustness of our model, we have retained our formulation, for reasons given in the text. We also thank the referee for pointing out that equations (5) and (6) can be generalized as

$$
\begin{aligned}
h_A\left(\frac{\dot{A}}{A}\right)^2 + h_B\left(\frac{\dot{B}}{B}\right)^2 + h_D\left(\frac{\dot{D}}{D}\right)^2 + h_E\left(\frac{\dot{E}}{E}\right)^2 &= h_0\left(\frac{EL_R}{K}\right)^{0.5} + h_1\left(\frac{L_R}{L}\right)^{0.5} \\
&= h_0\left(\frac{El_R}{k}\right)^{0.5} + h_1(l_R)^{0.5}.
\end{aligned}
\tag{7}
$$

Under the assumption of factor-augmenting technical progress, capital enters the production function in efficiency terms, AK. An economy can choose between increasing the number of physical units of inputs or their efficiency, and the decision will be determined by marginal considerations. The question then is how these marginal conditions are affected by the formulation of the innovation possibilities frontier. Below we will examine two possible specializations of the function given in equation (7) and their economic implications.

If $h_0 = 0$, then the position of the production possibilities frontier is determined by the allocation of labour for research, and we have an assumption that is comparable with the one in Uzawa (1965) and Lucus (1993). But it assumes that the position of the innovation possibility frontier is determined by a scalar, l_R, without involving the capital–labour ratio or the output per unit of labour, a form similar to investment functions in neoclassical growth models.[5] Second, while the efficiency of labour in production increases, the efficiency in research does not.

We therefore propose an alternate formulation which differs from the above in two respects. First, the efficiency of labour in all industries increases and the rate of increase in various types of employment is determined by the position on the production possibilities frontier. Second, the position is determined not only by l_R but also by k; in short, we assume that the more capital-intensive the economy is, the more resources are required to achieve proportional increase in efficiency. So we set $h_1 = 0$ and $h_0 \neq 0$. As we pointed out in footnote 4, the qualitative nature of the results remain the same whether we set $h_1 = 0$ or $h_0 = 0$.

Finally, we set $h_A = h_B = h_E = 1$.

The intertemporal optimization problem can now be stated as follows:

$$
\max \int_0^\infty e^{-\rho t}\frac{1}{L}F(AK, BL_c)dt,
$$

subject to (4), (6) and (7).

We find the following change of variables to be useful in solving this problem:

$$
g_1 \equiv \frac{AK}{\mathfrak{E}}, \quad g_2 \equiv \frac{BL}{\mathfrak{E}}, \quad g_3 \equiv \frac{DL}{K}, \quad \text{and } g_4 = \frac{EL}{K},
$$

$$
\begin{aligned}
C &= F(AK_C, BL_C), \\
\dot{K} &= V(K_K, DL_K) - \delta K,
\end{aligned}
$$

where $V(\cdot)$ is homogeneous of degree 1 in K_K and DL_K and it holds that $K = K_C + K_K$. As the referee pointed out, the steady-state results are not affected but the complexity of the dynamic analysis will increase; for this reason, we have not persued that suggestion in this paper.

[5] Note that in the Solow–Swan model $\dot{K}/K = sF(K, L)/K = sF(1, 1/k) = sf(1/k)$.

where $\mathfrak{E} = \mathfrak{E}_0 e^{\varepsilon t}$ and $\varepsilon =$ the growth rate of the system in steady state. After substitution (see Mathematical Appendix), the problem can be reformulated as

$$\max \int_0^\infty e^{-(\rho-\varepsilon+n)t} c\,dt, \tag{8}$$

subject to (4), (6) and (7). The Lagrangian is

$$\mathscr{L} = \mathrm{e}^{-\alpha t}\left\{ F(g_1, g_2 l_c) + \lambda_1\left(\frac{\dot{K}}{K} - l_K g_3 + \delta\right) + \lambda_2\left[\left(\frac{\dot{g}_1}{g_1} - \frac{\dot{K}}{K} + \varepsilon\right)^2 + \left(\frac{\dot{g}_2}{g_2} - n + \varepsilon\right)^2 \right.\right.$$
$$\left.\left. + h_D\left(\frac{\dot{g}_3}{g_3} + \frac{\dot{K}}{K} - n\right)^2 + \left(\frac{\dot{g}_4}{g_4} - n - \frac{\dot{K}}{K}\right)^2 - h_0(g_4 l_R)^{0.5}\right]\right\} = e^{-\alpha t}\mathscr{F}, \tag{9}$$

where $\alpha = \rho - \varepsilon + n = \rho - \beta^* > 0$, where $\beta^* = (\dot{B}/B)^*$ is the equilibrium rate of growth of efficiency of labour.

There are seven state variables: l_C, l_K, K, g_1, g_2, g_3 and g_4. Corresponding to each of these state variables is a Euler equation giving the intertemporal path of the variable:

$$\mathscr{L}_s - \frac{\mathrm{d}}{dt}\mathscr{L}_{\dot{s}} = 0,$$

where $s = l_C$, l_K, K, g_1, g_2, g_3 and g_4.

We shall now examine and interpret these equations.

1. $\mathscr{L}_s - \dfrac{d}{dt}\mathscr{L}_{\dot{s}} = 0$

Since the $\dot{l}_C$ do not appear in the Lagrangian, the second term, $(d/dt)\mathscr{L}_{\dot{l}_c} = 0$. Expanding the first term, we get

$$F_2 g_2 + \frac{1}{2}\lambda_2 h_0 g_4^{0.5} l_R^{-0.5} = 0. \tag{10}$$

If you transfer one unit of labour from research to the production of consumption goods, the increase in the per capita output of the consumption good will be $F_2 g_2/L$. The shift in the innovation possibility curve will lead to a decrease in the contribution of the researcher sector; when valued at the shadow price given by the Lagrangian multiplier λ_2, the economic value of reduced employment in research will be $-\frac{1}{2}\lambda_2 h_0 g_4^{0.5} l_R^{-0.5}(1/L)$. At the margin, these two should be equal, leading to the equation (10). Hence

$$\lambda_2 = -2\frac{F_2 g_2}{h_0 g_4^{0.5} l_R^{-0.5}} < 0. \tag{11}$$

2. $\mathscr{L}_{l_K} - (d/dt)\mathscr{L}_{\dot{l}_K} = 0$

Again, the second term of the equation is zero. Evaluating the first term, we get

$$-\lambda_1 g_3 + \frac{1}{2}\lambda_2 h_0 g_4^{0.5} l_R^{-0.5} = 0. \tag{12}$$

As before, we can consider the effect of transferring a unit of labour from research to the capital sector. The increased output of the capital sector (as it contributes to the production of consumption goods per capita in the future) is valued at $-\lambda_1(g_3/L)$. (Note that λ_i, $i = 1, 2$, is negative.) The economic value of a decrease in the research sector, as argued in the previous paragraph, is the negative of the second term on the left-hand side of (12) divided by L. Equating the two and substituting for λ_2, we get

$$\lambda_1 = -F_2 \frac{g_2}{g_3} < 0. \tag{13}$$

3. $\mathscr{L}_K - (d/dt)\mathscr{L}_{\dot{K}} = 0$

$$\begin{aligned} &\frac{\alpha\lambda_1}{K} - 2\alpha\frac{\lambda_2}{K}\left(\frac{\dot{A}}{A} - h_D\frac{\dot{D}}{D} - \frac{\dot{E}}{E}\right) - \frac{\dot{\lambda}_1}{K} + 2\frac{\dot{\lambda}_2}{K}\left(\frac{\dot{A}}{A} - h_D\frac{\dot{D}}{D} - \frac{\dot{E}}{E}\right) \\ &\quad + 2\frac{\lambda_2}{K}\left[\frac{\dot{A}}{A}\left(\frac{\ddot{A}}{\dot{A}} - \frac{\dot{A}}{A}\right) - h_D\frac{\dot{D}}{D}\left(\frac{\ddot{D}}{\dot{D}} - \frac{\dot{D}}{D}\right) - \frac{\dot{E}}{E}\left(\frac{\ddot{E}}{\dot{E}} - \frac{\dot{E}}{E}\right)\right] = 0. \end{aligned} \tag{14}$$

If we consider the steady state, the last three terms of (14) vanish. Since K is used only in the production of consumer goods, there are no allocational effects to consider as in the two previous cases. An increase in K, keeping g_1 constant, affects the Lagrangian only through the shift of the innovation possibility frontier. To determine this effect, note that, from (9), $\mathscr{L} = e^{-\alpha t}\mathscr{F}$; then $\mathscr{L}_K = e^{-\alpha t}\mathscr{F}_K$ and $(d/dt)\mathscr{L}_{\dot{K}} = -\alpha e^{-\alpha t}\mathscr{F}_{\dot{K}} + e^{-\alpha t}(d/dt)\mathscr{F}_{\dot{K}}$. In steady state, $\mathscr{F}_K = (d/dt)\mathscr{F}_{\dot{K}}$ and the two terms will cancel in the Euler equation, leaving only the term:

$$\alpha e^{-\alpha t}\mathscr{F}_{\dot{K}} = \frac{\alpha\lambda_1}{K} - 2\alpha\frac{\lambda_2}{K}\left(\frac{\dot{A}}{A} - h_D\frac{\dot{D}}{D} - \frac{\dot{E}}{E}\right).$$

Hence

$$\frac{\dot{A}}{A} - h_D\frac{\dot{D}}{D} - \frac{\dot{E}}{E} = \frac{1}{2}\frac{\lambda_1}{\lambda_2}. \tag{15}$$

4. $\mathscr{L}_{g_1} - (d/dt)\mathscr{L}_{\dot{g}_1} = 0$

$$F_1 + 2\alpha\frac{\lambda_2}{g_1}\left(\frac{\dot{A}}{A}\right) - 2\alpha\frac{\dot{\lambda}_2}{g_1}\left(\frac{\dot{A}}{A}\right) - \frac{2\lambda_2}{g_1}\frac{\dot{A}}{A}\left(\frac{\ddot{A}}{\dot{A}} - \frac{\dot{A}}{A}\right) = 0. \tag{16}$$

In steady state, the last two terms of (16) will vanish. The first term arises from the direct contribution of an increase in g_1 to output per person. The second term is $-\alpha f_{\dot{g}_1}$. From (16), we can derive the rate of growth of the efficiency of capital in the steady state:

$$\frac{\dot{A}}{A} = -\frac{1}{2}\left(\frac{F_1 g_1}{\alpha\lambda_2}\right) = \frac{1}{4}\frac{h_0}{\alpha}\frac{F_1 g_1}{F_2 g_2}\left(\frac{g_4}{l_R}\right)^{0.5} \tag{17}$$

The last expression is derived by substituting for λ_2 from (11).

5. $\mathscr{L}_{g_2} - (d/dt)\mathscr{L}_{\dot{g}_2} = 0$

$$F_2 l_c + 2\alpha \frac{\lambda_2}{g_2}\left(\frac{\dot{B}}{B}\right) - 2\frac{\dot{\lambda}_2}{g_2}\left(\frac{\dot{B}}{B}\right) - 2\frac{\lambda_2}{g_2}\left(\frac{\dot{B}}{B}\right)\left(\frac{\ddot{B}}{\dot{B}} - \frac{\dot{B}}{B}\right) = 0. \qquad \textbf{(18)}$$

As with the previous equation, the last two terms vanish in steady state. The first term represents the marginal product of g_2. The second term is $-\alpha \mathscr{F}_{g_2}$. Equation (18) can be used to derive the growth in the efficiency of labour in steady state:

$$\frac{\dot{B}}{B} = -\frac{1}{2}\frac{F_2 g_2 l_C}{\alpha \lambda_2} = \frac{1}{4}\frac{h_0 l_c}{\alpha}\left(\frac{g_4}{l_C}\right)^{0.5}. \qquad \textbf{(19)}$$

$$6.\ \mathscr{L}_{g_3} - d/dt\, \mathscr{L}_{\dot{g}_3} = 0 - \lambda_1 l_K + 2\alpha h_D \frac{\lambda_2}{g_3}\frac{\dot{D}}{D} - 2h_D \frac{\dot{\lambda}_2}{g_3}\left(\frac{\dot{D}}{D}\right)$$
$$-2\frac{h_D \lambda_2}{g_3}\left(\frac{\dot{D}}{D}\right)\left(\frac{\ddot{D}}{\dot{D}} - \frac{\dot{D}}{D}\right) = 0. \qquad \textbf{(20)}$$

The last two terms again vanish in steady state. The term g_3 is the state variable corresponding to the productivity of labour in the sector producing capital goods. It does not enter the production function for consumption goods and so there is no term corresponding to the marginal product in the two previous equations. The first term reflects the higher output of capital (note that λ_1 is negative), as the productivity of labour in the sector increases. The second term is, as in previous equations, $-\alpha \mathscr{F}_{g_3}$. Equation (20) gives the steady-state productivity growth of labour in the capital goods sector.

$$\frac{\dot{D}}{D} = \frac{1}{4}\frac{h_0}{\alpha h_D} g^{0.5} l_R^{-0.5} l_K. \qquad \textbf{(21)}$$

$$7.\ \mathscr{L}_{g_4} - (d/dt)\mathscr{L}_{\dot{g}_4} = 0$$

$$-\frac{1}{2}h_0\lambda_2\left(\frac{l_R}{g_4}\right)^{0.5} + 2\alpha\frac{\lambda_2}{g_4}\frac{\dot{E}}{E} - 2\frac{\dot{\lambda}_2}{g_4}\frac{\dot{E}}{E} - 2\frac{\lambda_2}{g_4}\left(\frac{\dot{E}}{E}\right)\left(\frac{\ddot{E}}{\dot{E}} - \frac{\dot{E}}{E}\right) = 0. \qquad \textbf{(22)}$$

The last two terms will vanish in steady state. An increase in g_4 will shift the innovation possibility frontier outwards and the value of that shift is given by the first term. The second term is $-\alpha \mathscr{L}_{\dot{g}_4}$. The growth rate of the efficiency of labour in research sector is

$$\frac{\dot{\mathrm{E}}}{\mathrm{E}} = \frac{1}{4}\frac{h_0}{\alpha}(g_4 l_R)^{0.5}. \qquad \textbf{(23)}$$

Equations (11), (13), (15), (17), (19), (21) and (23) define the steady state. *Hence this model has a steady state with positive growth rates for physical capital and positive rates of factor augmentation.*

In microeconomic models of endogenous technical progress, the firm is assumed to be a price-taker in the input markets. So the differential growth in prices is the motivating factor for the firm to choose more productive technology. In a growth theory, the input prices are determined as part of the general equilibrium. It is the factor endowment and the characterization of the technology that determine the path of input prices and resource allocation.[6]

Samuelson argued that, with a stationary innovation possibility frontier, the system will converge to Harrod-neutral technical progress. Further, the bias at any point of time is determined by the ratio of factor shares; in other words, the tangent to the innovation possibility frontier has a slope equal to the ratio of factor shares.

We have assumed in this model that consumer goods are produced with labour and capital while the other two sectors have only one input, labour. The innovation possibility frontier is not stationary but shifts with the allocation of resources to research, as in the Nordhaus (1967), Kamien–Schwartz (1969) and Sato–Ramachandran (1987) models. So at any point in time the position of the innovation possibility frontier and the position of the economy along that frontier have to be determined. In this paper, therefore, we have introduced a Hicksian hypothesis that bias is determined by the scarcity of inputs (reflected in microeconomic models by their prices) in a general equilibrium model with a Kennedy–Weizäcker–Samuelson innovation possibility frontier, and we have shown that the model can have a steady-state with bias.

The bias in technical progress in the consumer sector, in steady state, can be determined from the (17) and (19):

$$\frac{\dot{A}/A}{\dot{B}/B} = \frac{g_1 F_1}{g_2 l_C F_2}. \tag{24}$$

The right-hand side of (24) equals the ratio of the income of the two inputs in the consumer sector; if we accept the stylized fact that labour's share exceeds that of capital, then $\dot{A}/A < \dot{B}/B$. This is shown in Figure 1.

If r is the rental of capital and w is the wage rate, the national income is $rK + wL$. Consumption equals the value of consumer goods produced: $rK + wL_C$. So the savings in the economy is $w(L_K + L_R)$. As in any endogenous growth model, this savings is invested to increase the physical units of capital and to increase the efficiency of inputs.

The derived bias in this model agrees with Sato (1970), which estimated that growth in the efficiency of labour exceeds that of capital though both are positive. But Sato (1970) assumed that the economy produces one malleable good. In this model the economy is divided into three sectors, and the maintained assumption is that the growth of productivity of labour in the capital goods sector is less than that in the consumption goods sector. Kendrick and Grossman (1980) estimate the productivity growth in the private domestic business economy and in specific sectors. The growth rate of total factor productivity from 1948 to 1977 is 2.3%, while total factor productivity in the non-electrical machinery sector grew at a rate of 1.07%. Productivity in the electrical machinery industry grew at 3.7% (Kendrick and Grossman, 1980, pp. 114, 156–157).[7] The

[6] The sufficiency condition can be obtained by evaluating the principal minors of the Hessian, H, at the steady state. The first principal minor $|H_1| = (0.5)(1 - 0.5)h_0\lambda_2(g_4)^{0.5}(l_R)^{-1.5} < 0$ as λ_2 is negative. All other minors are equal to zero, either because of the coefficient for the parameters chosen or because all the elements of one of the rows or columns are equal to zero. This shows that the Lagrangian is quasi-concave.

[7] See Jorgenson (1990) for detailed estimates based on a different set of assumptions.

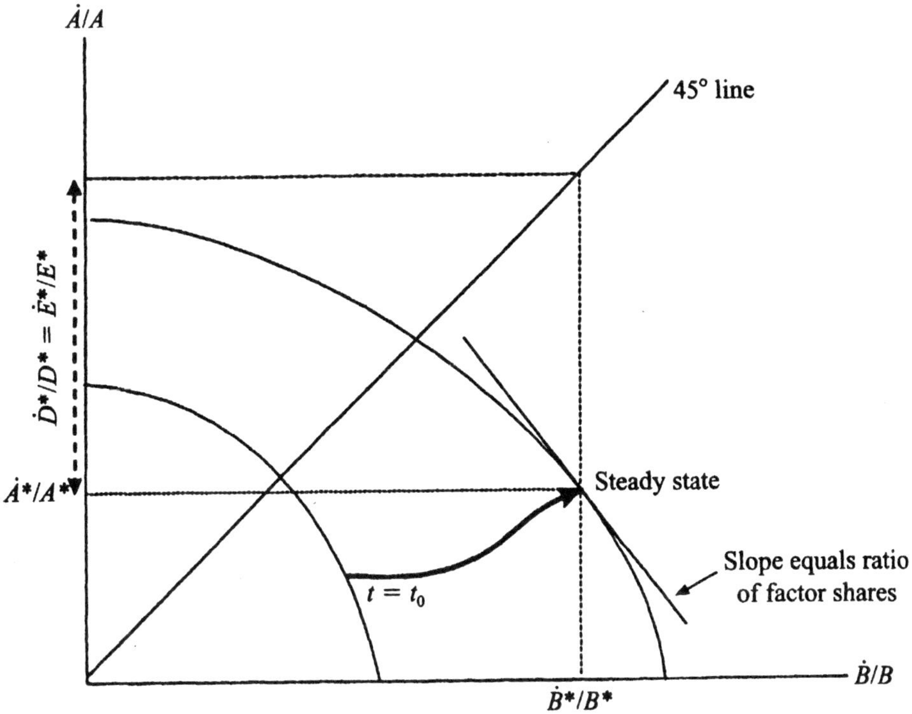

Figure 1. The Innovation Possibility Frontier and Transition to the Steady State.

higher growth of the productivity of the business sector in general can be taken as a preliminary indication of differential growth in the consumption sector compared with the capital goods sector.

4. Outside The Steady State

A number of indicators show that the economy is not in steady state. Calculating the output–capital ratio from the data for the private domestic economy from Kendrick and Grossman (1980, p. 114; quantities are normalized to 1967 = 100), the ratio increases from 0.772 to 1.057. Nor can this increase be attributed to increases in the efficiency of capital, making Y/AK constant; if that were true then AK/BL would be constant, but Table 1 shows that it has decreased from 1910 to 1960. The transitional time required to convergence can be very large, as was noted in Sato (1963).

Analysis of an optimal model along an adjustment path is even more difficult than that of the steady state. One solution is to see whether the system satisfies some "conservation laws", defined as scalar functions that are constant along the optimal trajectory. The advantage for empirical analysis is that, if conservation laws can be derived from the model, then its verification provides a relatively easy test for the model itself. In economics, Samuelson (1970) derived the constancy of the income–wealth ratio as the conservation law for the von

Neumann model. Sato (1981) derived many other conservation laws using Lie groups, and these ratios for OECD countries are examined in Sato (1997).

For this model, the conservation law is that the ratio

$$\alpha = \frac{F(g_1, g_2 l_c) + F_2 g_2 l_K\left(1 - \dfrac{\delta}{g_3 l_K}\right) + 4F_2 g_2 l_R\left[1 - \dfrac{n\left(\dfrac{\dot{A}}{A} - h_D\dfrac{\dot{D}}{D} - \dfrac{\dot{E}}{E}\right) - (n-\varepsilon)\left(\dfrac{\dot{A}}{A} + \dfrac{\dot{B}}{B}\right)}{h_0 g_4^{0.5} l_R^{0.5}}\right]}{\displaystyle\int_t^{\infty} e^{-\alpha(s-t)} F(g_1(s), g_2(s) l_C(s))\, ds}$$

The denominator is the present value of the future consumption stream and can be identified as a measure of wealth. The numerator is the per capita output of consumption goods at time t together with terms reflecting the income earned in the intermediate sectors. The possibility of using this relation in empirical work remains to be explored.

5. Conclusion

If productivity growth is to be achieved only by the allocation of scarce resources to research, the question raised earlier of whether or not it is efficient to allocate some resources to increasing the efficiency of capital cannot be avoided. In microeconomic models, where the firm is taken as price-taker in the input market, the relative growth of input prices can be taken as the driving force. In macroeconomic models, the relative growth in input prices has to be derived from the model itself in terms of the temporal paths of resources and technology. Formulating a model to address the question of optimal bias poses many problems. In Sato *et al.* (1999) and in this paper, we have attempted to develop tractable models to analyse the optimal intertemporal allocation of resources leading to a steady-state bias. These papers reveal the crucial role of the technology of the production and research sectors in determining the allocation of resources for physical capital accumulation and for enhancing the productivity of inputs. Our understanding of these processes will determine the answer to the allocation question.

Mathematical Appendix

The following transformation of variables were made to facilitate the definition of the steady state.

$$g_1 \equiv \frac{AK}{\epsilon},\ g_2 \equiv \frac{BL}{\epsilon},\ g_3 \equiv \frac{DL}{K},\ g_4 \equiv \frac{EL}{K},$$

where A, B, D, E, K and L are as defined in the text and the rate of change of is the growth rate in steady state, ε. Further,

$$l_C \equiv \frac{L_c}{L}, \quad l_K \equiv \frac{L_K}{L}, \quad \text{and } l_R \equiv \frac{L_R}{L}; \quad l_C + l_K + l_R = 1.$$

Hence,

$$\frac{\dot{g}_1}{g_1} = \frac{\dot{A}}{A} + \frac{\dot{K}}{K} - \varepsilon; \quad \frac{\dot{g}_2}{g_2} = \frac{\dot{B}}{B} + n - \varepsilon; \quad \frac{\dot{g}_3}{g_3} = \frac{\dot{D}}{D} + n - \frac{\dot{K}}{K} \quad \text{and } \frac{\dot{g}_4}{g_4} = \frac{\dot{E}}{E} + n - \frac{\dot{K}}{K},$$

$$c = \frac{C}{L} = \frac{c}{L} F\left(\frac{AK}{\epsilon}, \frac{BL}{\epsilon}\frac{L_C}{L}\right) = e^{(\varepsilon - n)t} F(g_1, g_2 l_C),$$

$$\frac{\dot{K}}{K} = \frac{DL}{K}\frac{L_K}{L} - \delta = g_3 l_K - \delta, \tag{A1}$$

and

$$\left(\frac{\dot{A}}{A}\right)^2 + \left(\frac{\dot{B}}{B}\right)^2 + h_D\left(\frac{\dot{D}}{D}\right)^2 + \left(\frac{\dot{E}}{E}\right)^2 = h_0\left(\frac{EL}{K}\frac{L_R}{L}\right)^2 = h_0(g_4 l_R)^{0.5} \tag{A2}$$

The problem is then to maximize

$$\max \int_0^\infty e^{-\alpha t} c dt$$

subject to the constraints given by the (A1) and (A2). This leads to the Lagrangian in the text.

Final version accepted 9 April 1998.

References

Drandakis, E. M. and E. S. Phelps (1966) "A Model of Induced Invention, Growth and Distribution", *Economic Journal*, Vol. 76, pp. 823–840.

Grossman, G. M. and E. Helpman (1993) "Endogenous Innovation in the Theory of Growth", *Journal of Economic Perspectives*, Vol. 8, pp. 23–44.

Hicks, J. R. (1932) *Theory of Wages*, London: Macmillan.

Jorgenson, D. W. (1990) "Productivity and Growth", in E. R. Berndt and J. E. Triplet (eds.), *Fifty Years of Economic Measurement: The Jubilee of the Conference on Research in Income and Wealth*, Chicago: University of Chicago Press.

Kamien, M. I. and N. L. Schwartz (1969) "Induced Factor Augmenting Technical Progress for a Microeconomic Viewpoint", *Econometrica*, Vol. 37, pp. 668–684.

Kendrick, J. W. and E. S. Grossman (1980) *Productivity in the United States: Trends and Cycles*, Baltimore: Johns Hopkins University Press.

Kennedy, C. (1964) "Induced Bias in Innovation and the Theory of Distribution", *Economic Journal*, Vol. 74, pp. 541–547.

Lucus, R. E. Jr (1993) "Making a miracle", *Econometrica*, Vol. 61, pp. 251–272.

Nordhaus, W. D. (1967) *Invention, Growth and Welfare*, Cambridge, Mass.: MIT Press.

Salter, W. E. G. (1960) *Productivity and Technical Change*, Cambridge: Cambridge University Press.

Samuelson, P. A. (1965) "A Theory of Induced Innovation along Kennedy-Weizäcker Lines", *Review of Economics and Statistics*, Vol. 47, pp. 343–356.

Samuelson, P. A. (1970) "Law of Conservation of the Capital–Output Ratio", *Proceedings of the National Academy of Science, Applied Mathematical Science*, pp. 1477–1479.

—— (1983) "Thünen at Two Hundred", *Journal of Economic Literature*, Vol. 21, pp. 1468–1488.

Sato, R. (1963) "Fiscal Policy in a Neoclassical Growth Model: An Analysis of the Time Required for Equilibrating Adjustment", *Review of Economic Studies*, Vol. 30, pp. 16–20; reprinted in *Growth Theory and Technical Change: The Selected Essays of Ryuzo Sato*, Vol. 1, Cheltenham: Edward Elgar, 1996.

—— (1970) "The Estimation of Biased Technical Progress and the Production Function", *International Economic Review*, Vol. 11, pp. 179–208; reprinted in *Growth Theory and Technical Change: The Selected Essays of Ryuzo Sato*, Vol. 1, Cheltenham: Edward Elgar, 1996.

—— (1981) *Theory of Technical Change and Economic Invariance*, New York: Academic Press; reprinted with revisions: Cheltenham: Edward Elgar, 1999.

—— (1996) "A Note on Modelling Endogenous Growth", *Keio Economic Studies*, Vol. 33, pp. 93–101.

—— (1997) "Optimal Economic Growth: Test of Income–Wealth Conservation Laws in OECD Countries", working paper, Center for Japan–US Business and Economic Studies, New York University.

—— and R. V. Ramachandran (1987) "Factor Price Variation and the Hicksian Hypothesis", *Oxford Economic Papers*, Vol. 39, pp. 343–356; reprinted in *Growth Theory and Technical Change: The Selected Essays of Ryuzo Sato*, Vol. 1, Cheltenham: Edward Elgar, 1996.

——, R. V. Ramachandran and C. Lian (1999) "A Model of Optimal Economic Growth with Endogenous Bias", *Macroeconomic Dynamics*, Vol. 3, pp. 293–310.

Solow, R. M. (1956) "A Contribution to the Theory of Economic Growth", *Quarterly Journal of Economics*, Vol. 70, pp. 65–94.

—— (1957) "Technical Change and Aggregate Production Function", *Review of Economics and Statistics*, Vol. 39, pp. 312–320.

Uzawa, H. (1965) "Optimal Technical Change in an Aggregative Model of Economic Growth", *International Economic Review*, Vol. 6, pp. 18–31.

Chapter 5

ESTIMATION OF BIASED TECHNICAL PROGRESS

Empirical work on neoclassical growth models led to the recognition that technological progress is the dominant factor in the growth of *per capita* income. This led the economic profession to explore four questions: (i) How important is technological and technical progress in the process of economic growth? (ii) What is the cause of technical progress - is it exogenous or endogenous to the economic system? (iii) How is technological change transmitted into technical progress in the macroeconomy? and (iv) If technical progress can be classified as labor saving, neutral or capital saving, is there any systematic bias in an economy towards any particular kind of technical progress and, if so, why? (Jones, pp.154–5).

In the subsequent discussion of these questions, the contribution of Hicks (1932) was definitive. Well before the advent of modern growth theory, he proposed a classification of technical progress and stated the conditions under which technical progress can be labor saving, neutral or capital savings. It was seen that the effect of technical progress on share of income going to one of the factors will depend on the nature of Hicksian bias. The debate over the relative merits of this classification over that proposed by Harrod has led to a clear understanding of how to represent factor-augmenting technical progress in models with a neo-classical production function.

While Hicks was concerned with micro-economic questions, the stability of the share of wages in the economy as a whole needed explanation. The innovation possibilities frontier which determined the degrees of bias that are technologically feasible was offered as an explanation. The works of Kennedy, Samuelson, and von Wiezsäker on macromodels and of Ahmed, Nordhaus, Kamien and Scwatrz, and Sato and Ramachandran on micromodels explored the possibilities of developing a theory of induced technical progress. This theory ran into two road blocks. First, at a theoretical level, growth models seem to have steady state only if technical progress is Harrod-neutral; while Harrod neutral technical progress is a type of biased technical progress in terms of Hicksian classification, it would be trivial to have a theory of biased technical progress if it admits only one specific type of bias. Second, econometric efforts to measure the degree of bias ran into the Diamond-McFadden impossibility theorem.

So the profession became skeptical of this approach and recently it is claimed that this is one of the lines of research that has reached a dead end [Ruttan (1997, p.1520)]. It has been argued that "if economics really wants to take technology seriously, economics will have to become a more historical discipline" [Wright 91997, p. 1565)]. In this empirical paper and in a related series of theoretical papers (Sato and Ramachandran (1988), Sato, Ramachandran and Lian (1988), we argue that studies in the Hicksian tradition of biased technical progress can lead to meaningful results.

1. Historical Evidence for Biased Technical progress

The lead England had over the continent in industrial revolution during the eighteenth century and the impressive degree of mechanization in America in the nineteenth century are attributed by economic historians to the incentive for faster innovation and to adaptations arising from the scarcity of resources. Mantoux (1928) explained the process of innovation in the eighteenth century as generated by pressures to develop certain types of innovations. Rothbarth (1946) argued that the scarcity of labor in U.S. encouraged mechanization and Habakkuk (1962) provides an authoritative analysis of the historical evidence affecting mechanization in England and the United States. Even though these studies did not use quantitative analysis, they are worth considering for the insights they provide, given that econometric studies have their own measurement problems.

Habakkuk (pp. 4–5) observes that the rate of adoption of technical innovation is conditioned by three types of influences. The first one is sociological and includes the educational system and the character of the entrepreneurs. The second is the rate of capital accumulation which determines the opportunities to install new technologies. The third influence is the scarcities that capital accumulation creates; if capital grows faster than cooperating factors, it first effect is to reduce the rate of profit. The falling profit rates encourage search for new methods "in any direction" and, as some factors become scarcer sooner, it encourages innovation "in a particular direction". He also recognized that reduction in profits may reduce rate of investment so that the second and third influences may go in opposite directions.

"Why should mechanization, standardization and mass-production have appeared *before* 1850 and to an extent which surprised reasonably dispassionate English observers?", he asks (p. 5). The second influence, the higher rate of investment in the United States as compared to England, is ruled out as the reason for it. Though many studies indicate that labor commands relatively higher wages in the States, the question remains why did English entrepreneurs not replace inexpensive English labor with machines produced at low cost by these laborers? (p. 8). Should not the free availability of land and higher wages restrict the development of industry in the States as many writers including Adam Smith predicted? Habakkuk also recognizes that scarcity of labor does not necessarily make labor saving innovations more feasible or profitable.

Industries began to emerge in the United States at a time when wages were substantially higher than in England and the supply of labor was less elastic. The productivity of the American agriculture and the prevalence of family farms where the total earnings were divided equally among members, meant that firms had to pay wages equal to the high average product of U.S. agriculture while in England it need equal only the marginal product (p. 14). The differential for skills was smaller in the US than in England (p. 20); a lower spread in wages make it cheaper to used skilled labor with machines than unskilled labor.

Adoption of innovations depended on the available spectrum of techniques; however, in those years when most of the innovations came from engineers tinkering in the workplace than scientists working in research divisions or laboratories, the distinction between innovation and adoption was even less clear than today. Some of the innovations in the eighteenth century were such that they were manifestly superior at all input prices but the profitability of many others including power loom, depended on the relative input prices. Prices of land and energy also affected the adoption of new techniques. The adoption of water looms was more noticeable in Massachusetts, where energy was cheaper than in Rhode Island, where it was scarce (p. 33).

To have an assessment of the technological and economic possibilities visualized by the innovators in the eighteenth century, MacLeod (1988) examines the stated goals of those who filed for patent between 1660 and 1799. Patentees throughout the period placed greater emphasis on improving the quality of the product and in saving capital. It is true that one must accept the professions with a grain of salt as it was politically unwise to claim labor saving when "an invention fell under the disqualification of 'inconvenient' if its implementation would result in the displacement of workers." (*ibid*, p. 161). But there is no evidence that patents were rejected on that ground. By 1790's the resistance to labor savings innovations had reduced considerably and the percentage of innovations that claimed to save labor increased considerably. Yet only one in 5 patentees stated labor saving as the primary objective while half as much claimed to be capital saving. The constancy of the capital labor ratio in England from mid-eighteenth century to mid nineteenth century also supports that labor savings did not predominate.

However capital saving in the early years of industrialization alluded to savings of working capital rather than fixed capital. The interest payment to creditors who supplied raw material were substantial, and savings on inventory by speeding up the process resulted in substantial savings to the manufacturer. Mechanization in those days also contributed considerably to the reduction of waste due to unreliablity of the process, and many innovations concentrated on improving what we now would call quality control. Reducing the reliance on natural but uncertain sources of energy is shown by the fascination with perpetual motion. "In summary, then, there was growing interest throughout the eighteenth century, if not before, in labor-saving technology. Even accounting, however, for unwillingness among patentees to state intentions that appeared to threaten employment, it was always subordinate to the desire to save capital. Not surprisingly in an economy whose fixed assets were simple and cheap, the focus was almost entirely on working capital." (*ibid,* p. 180).

Mokyr (1990) takes an eclectic view and lists a number of factors but, on the whole, stresses factors determining the supply of innovations.

The cost of inputs had indeed an effect on innovations and their adoptions but labor savings was not the only concern that entrepreneurs had.

2. Theory of Endogenous Technical Progress

The ease of developing new technologies, the cost benefits of introducing them and the motive for the management to adopt them are analyzed in detail in various theories of endogenous technological progress. Hicks (1932) argued that "a change in the relative prices of factors of production is itself a spur to invention, and to invention of a particular kind - directed to economizing the use of a factor which has become relatively expensive".

Salter (1960), as is well known, objected to this line of argument claiming that what firms should be interested is in reducing total cost and not cost of a particular type. This criticism brought forth the idea of an innovation possibility frontier by Kennendy (1964) who used it also to develop a theory of distribution not based on marginal productivity of inputs. The innovation possibility frontier was integrated into the neoclassical theory of growth by Drandakis and Phelps (1966), von Weizäcker (1966) and Samuelson (1965). Sato and Ramachandran (1987) constructed a model where a profit maximizing monopolist will invest in technologies that increase capital and labor productivity at rates that depend on the growth of their prices.

Two questions were raised by this literature. The first one is whether the biased technical progress implies that only the productivity of labor increases or whether productivity of both increases but at different rates. Habakkuk emphasized the growth in productivity of labor though he did recognize that there are inducements to save other factors. MacLeod has down-played the role of labor savings in the eighteenth century England. The papers by Romer (1986) and Lucus (1993) has inspired a number of studies that attribute all productivity growth to the accumulation of human capital. So the first question is whether there is any empirical justification for being concerned with capital augmentation at all.

This brings up the second question: how can different growth rates of productivity of inputs be measured? The Diamond-McFadden impossibility theorem states that we have to have prior knowledge of the elasticity of substitution before estimating the growth rates in productivity of capital and labor separately. Sato (1970) showed one method to get around this limitation and in this paper, we use it to develop estimates of the rates of growth of productivity.

3. A Program for Measuring Biased Technical Progress

The production function with factor augmenting technical progress can be written as

$$Y = F[A(t)K(t), B(t)L(t)] \tag{1}$$

where $Y(t)$ is the output at time t; $K(t)$, the capital; $L(t)$, labor; and $A(t)$ and $B(t)$ are the "efficiencies" of $K(t)$ and $L(t)$ respectively.

Totally differentiating the production function, we get:

$$\frac{\dot{y}}{y} = \alpha\frac{\dot{A}}{A} + \beta\frac{\dot{B}}{B} + \alpha\frac{\dot{k}}{k} \tag{2}$$

where $y = Y/L, \alpha = (\partial F/\partial K)(K/Y)$, $\beta = (\partial F/\partial L)(L/Y) = 1 - \alpha$ and $k = K/L$.

The equation contains two unknowns, $\dot{A}/A$ and $\dot{B}/B$. We need an additional independent equation to estimate the two unknowns.

If technical progress is non-neutral, the value of the elasticity itself is influenced by the efficiencies of capital and labor. So Sato (1970, p. 182) introduced a new a modification of the definition of elasticity:

$$\sigma = \frac{d\left(\dfrac{AK}{BL}\right) \Big/ \dfrac{AK}{BL}}{d\left(\dfrac{\partial F/\partial BL}{\partial F/\partial AK}\right) \Big/ \dfrac{\partial F/\partial BL}{\partial F/\partial AK}} \tag{3}$$

Equating the marginal product of each input to its market price and differentiating with respect to time, we can derive the following relationships:

$$\frac{\dot{w}}{w} = \frac{\dot{B}}{B} - \frac{\alpha}{\sigma}\left(\frac{\dot{B}}{B} - \frac{\dot{A}}{A} - \frac{\dot{k}}{k}\right) \tag{4}$$

and

$$\frac{\dot{r}}{r} = \frac{\dot{A}}{A} + \frac{\beta}{\sigma}\left(\frac{\dot{B}}{B} - \frac{\dot{A}}{A} - \frac{\dot{k}}{k}\right) \tag{5}$$

The three equations for growth rate in per capita output, wage rate and rent, allow us to derive the following equations (when $\sigma \neq 1$):

$$\frac{\dot{A}}{A} = \frac{\sigma\dfrac{\dot{r}}{r} - \dfrac{\dot{Y}}{Y} + \dfrac{\dot{K}}{K}}{\sigma - 1} \tag{6}$$

and

$$\frac{\dot{B}}{B} = \frac{\sigma\dfrac{\dot{w}}{w} - \dfrac{\dot{Y}}{Y} + \dfrac{\dot{L}}{L}}{\sigma - 1} \tag{7}$$

Even now we have only two equations to determine three variables. σ, $\dot{A}/A$ and $\dot{B}/B$. This is the crux of the Diamond-McFadden impossibility theorem. One method around this problem is to estimate the elasticity of substitution first and then estimate the bias. Sato (1970) estimated the growth rates in productivity of capital and labor using CES and CEDD (Constant Elasticity of Derived Demand) production functions. Two results stand out. The productivity of capital grew at a positive rate but less than one percent per annum while labor productivity grew close to two percent per annum. If the years 1942 to 1945 are excluded, then there is a dramatic decline in the average growth rate of the efficiency of capital while growth rate of labor productivity increased over the previous estimate only slightly; given the disruptions of war economy, the result has to be interpreted with caution.

In this paper, we compare the growth of factor productivities in the United States and Japan using OECD data.

4. New Estimates of Productivity Growth in United States and Japan

United States

We begin by estimating the productivity growth for U.S. for the years 1909 to 1993 (see appendix for data sources). As a first step, we estimate average elasticity of substitution assuming Hicks neutral technical progress. For a year t, it is estimated using

$$\sigma_t^N = \frac{\frac{\dot{K}}{K} - \frac{\dot{w}}{w}}{\frac{\dot{w}}{w} - \frac{\dot{r}}{r}}. \tag{8}$$

We then estimate the average elasticity of substitution for the period as

$$\sigma_{AVG} = \frac{1}{T}\sum_{t=0}^{t=T} \sigma_t^N. \tag{9}$$

Next we plug this estimates of elasticity into the equations (6) and (7) for growth in factor efficiencies to obtain the estimate for each year.

If this value is the true value, then the average growth rates of efficiencies is as in Table 1.

Figure (1A&1B) plots the growth rates of efficiencies in various years.

Another way to test the results is to assume a CES production function with an elasticity estimated above. Taking the initial values of capital and labor and the assuming Hicks neutral technical progress, estimate the income for different years, Y^1, and plot it against the actual income. Repeat the exercise this time estimating income, Y^2, assuming not Hicks neutral technical progress but the

TABLE 1. Averages Assuming the Value of an Average Elasticity of Substitution.

σ_{AVG}	$\dot{A}/A$	$\dot{B}/B$
0.436	0.004	0.02

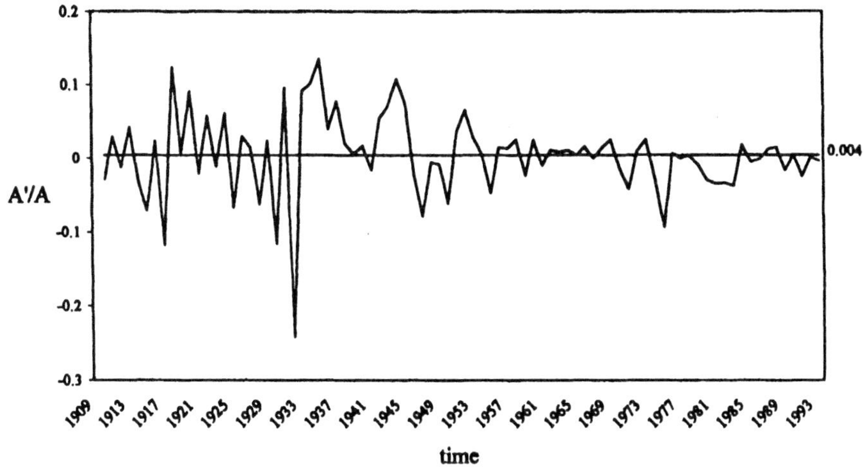

FIGURE 1A. Growth Rate of Capital Augmenting Technical Progress.

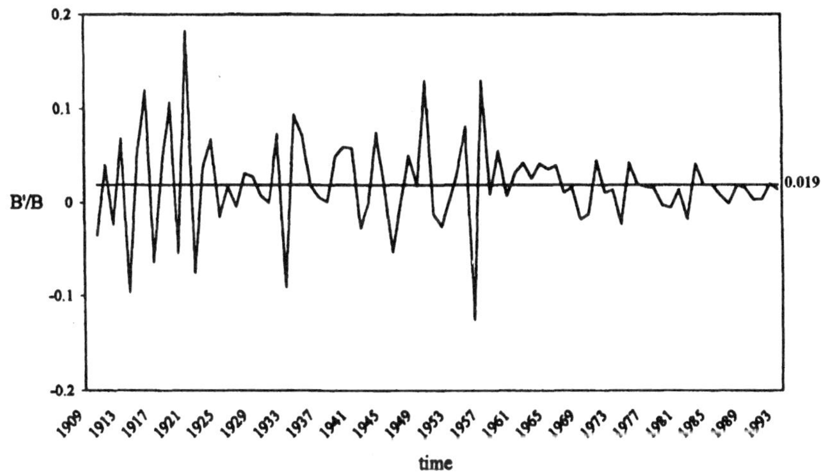

FIGURE 1B. Growth Rate of Labor Augmenting Technical Progress.

average growth rate in the efficiencies of the two inputs. Figure 2 shows that the plot of estimated income assuming biased technical progress tracks the actual income better than the one assuming Hicks neutral one.

But the estimate of the elasticity based on the assumptions Hicks neutrality has a bias if technical progress is indeed biased. If the efficiency of labor grows faster than that of capital and if the true value of elasticity is less than one, then the assumption of Hicks neutrality will bias it upwards. For

$$\frac{\partial \sigma^N}{\partial\left(\frac{\dot{B}}{B}-\frac{\dot{A}}{A}\right)} = -\frac{\left(\frac{\dot{w}}{w}-\frac{\dot{r}}{r}\right)(1-\sigma^N)}{\left(\frac{\dot{B}}{B}-\frac{\dot{A}}{A}+\frac{\dot{w}}{w}-\frac{\dot{r}}{r}\right)^2} < 0$$

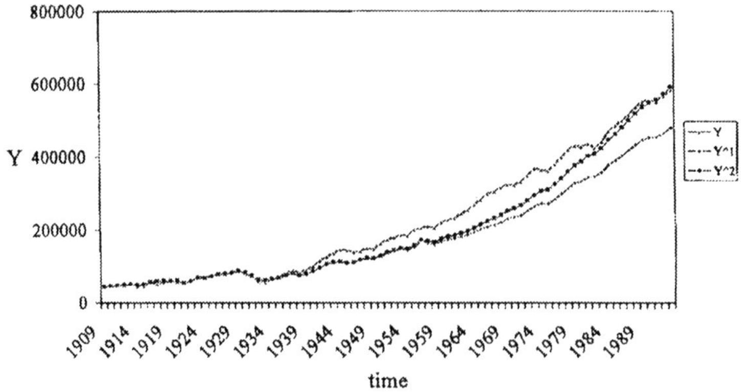

FIGURE 2. Actual vs. Estimated Y (Sigma = 0.502).

Given that the growth rate of wages exceed that of rent and that the number of independent studies indicate that elasticity of substitution is less than 1, we take 0.7 as the upper bound of the elasticity. To test the robustness of the results, we estimate the growth rates for different elasticities of substitution from 0.3 to 0.7. The average values are given in the table below.

TABLE 2. Average Growth Rates for Different Elasticities.

σ	$\dot{A}/A$	$\dot{B}/B$
0.4	0.004	0.020
0.6	0.004	0.019
0.8	0.004	0.020

What is striking about these numbers is that the growth rate of the efficiency of labor is consistently four to five times that of capital.

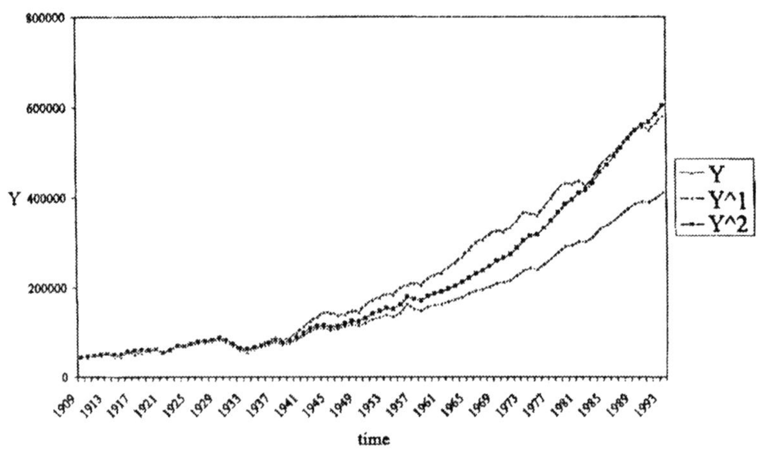

FIGURE 3. Actual vs. Estimated Y (Sigma = 0.4).

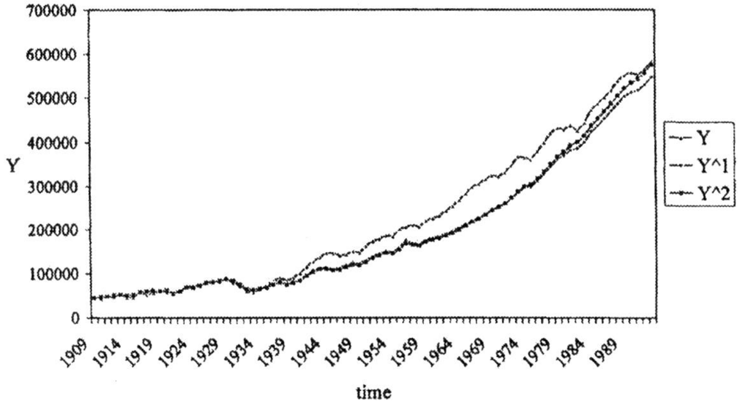

FIGURE 4. Actual vs. Estimated Y (Sigma 0.6).

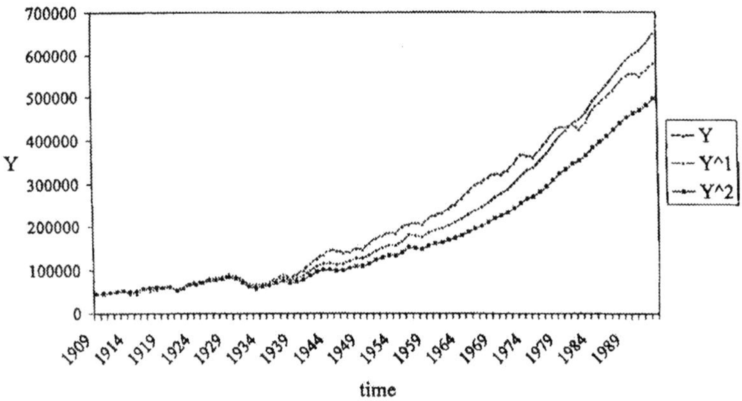

FIGURE 5. Actual vs. Estimated Y (Sigma = 0.8).

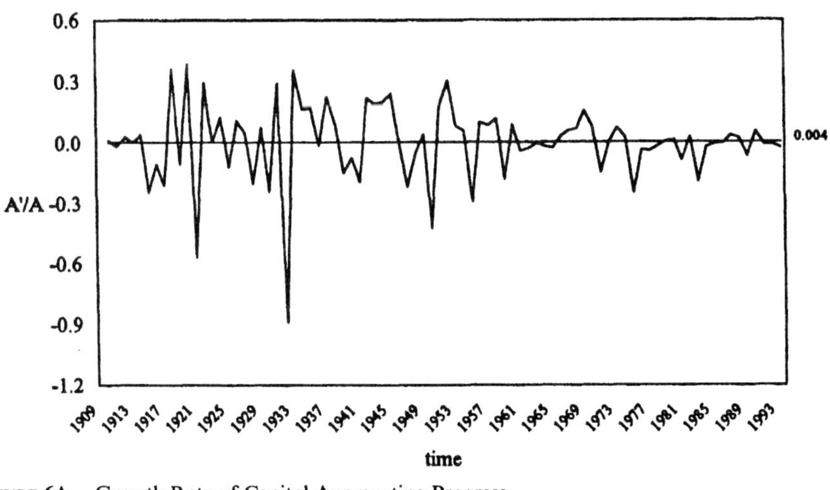

FIGURE 6A. Growth Rate of Capital Augmenting Progress.

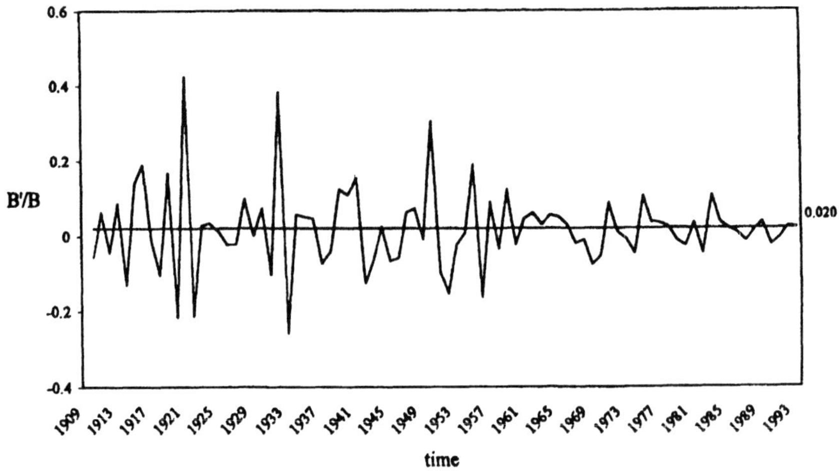

FIGURE 6B. Growth Rate of Labor Augmenting Technical Progress.

Figure 3, 4 and 5 plot actual and estimated income for elasticities 0.4, 0.6 and 0.8 while Figure 6 plots capital and labor efficiency over time with $\sigma = 0.8$.

Japan

The same method is used to estimate the growth rates in efficiencies in Japan. Unfortunately, there is considerable variation in the relative magnitudes of the growth rates of the two efficiencies in Japan. In fact, for elasticities of substitution greater than 0.8, $\dot{A}/A$ is positive and greater than $\dot{B}/B$ while for lower values of elasticity, $\dot{A}/A$ is not only less than $\dot{B}/B$ but turns negative. This is in direct contrast to the United States where the growth rate of A decreases as elasticity increases. Some of this can be explained by the shorter time series and for the inefficiencies of Japanese agriculture.

We use the same techniques as we did for the United States. First we calculate the average elasticity of substitution assuming Hicks neutral technical progress and then determine the growth rates in input efficiency.

TABLE 3. Average Growth Rates in Efficiency Assuming Average Elasticity of Substitution.

σ	$\dot{A}/A$	$\dot{B}/B$
0.902	0.037	0.019

As was done in the case of the United States, Figure 7 plots the growth rates of efficiency assuming the elasticity of substitution of 0.9. Figure 8 plots the actual and estimated income for the plot of actual income for Japan.

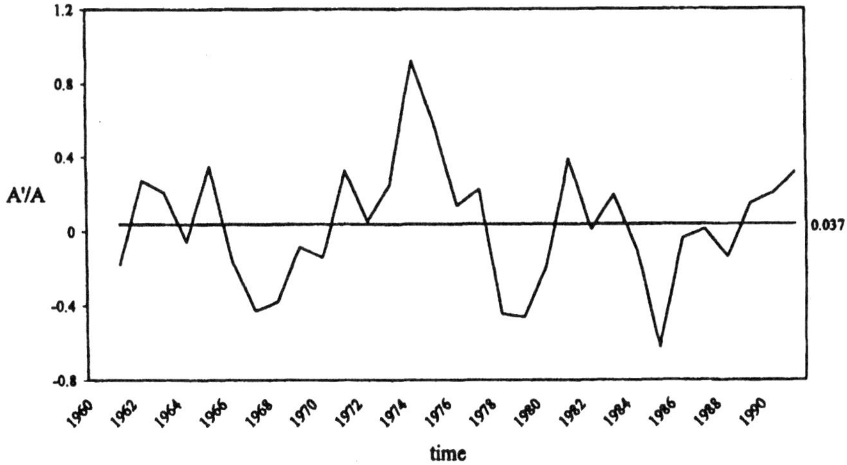

FIGURE 7A. Growth Rate of Capital Augmenting Technical Progress.

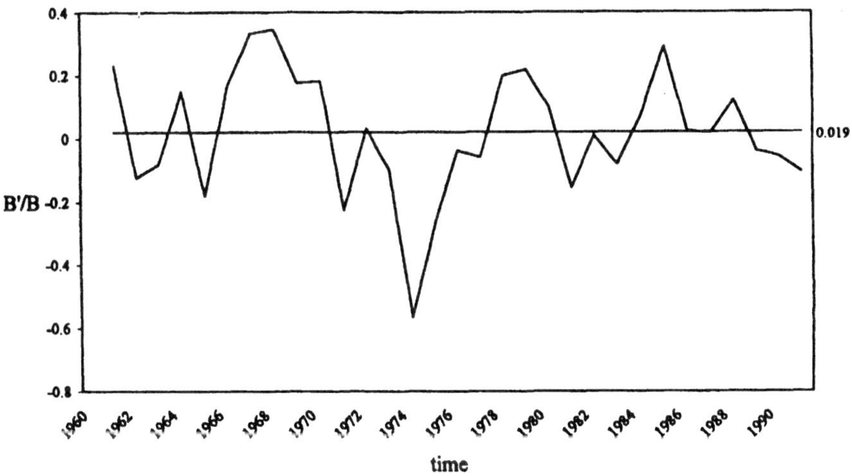

FIGURE 7B. Growth Rate of Labor Augmenting Technical Progress.

As noted in the discussion of the US data, the elasticities of substitution given above has a bias if the technological change has a bias. Using the same technique as we did there we recalculate the growth rates for efficiencies assuming different elasticities of substitution.

TABLE 4. Average growth rates for different elasticities.

σ	$\dot{A}/A$	$\dot{B}/B$
0.8	0.001	0.04
0.6	−0.016	0.051
0.5	0.02	0.053

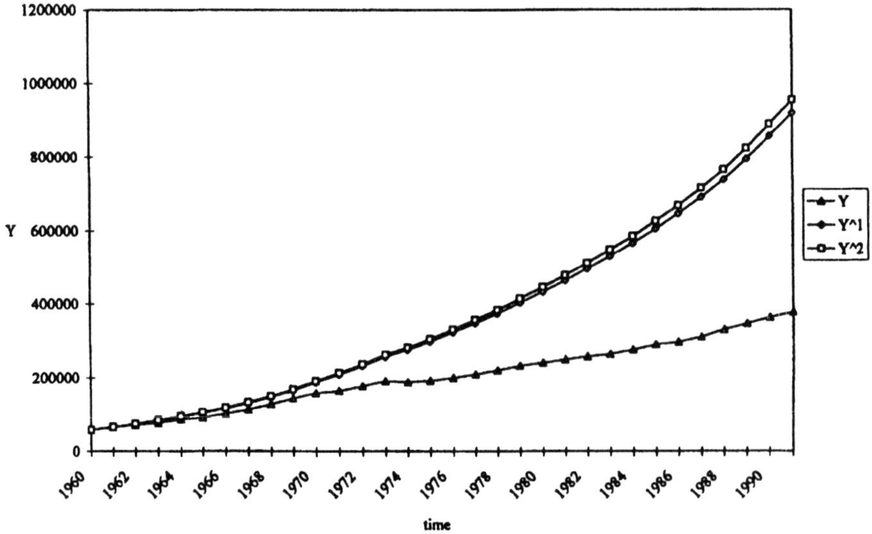

FIGURE 8. Actual vs. Estimated Y (Sigma = 0.902).

Figures 9 and 10 shows the trend in actual and estimated Y for elasticities of substitution 0.8 and 0.6. Figure 11 shows the growth rate of productivity for $\sigma = 0.6$. Figure 12 gives the values of estimated income and Figure 13 the trends

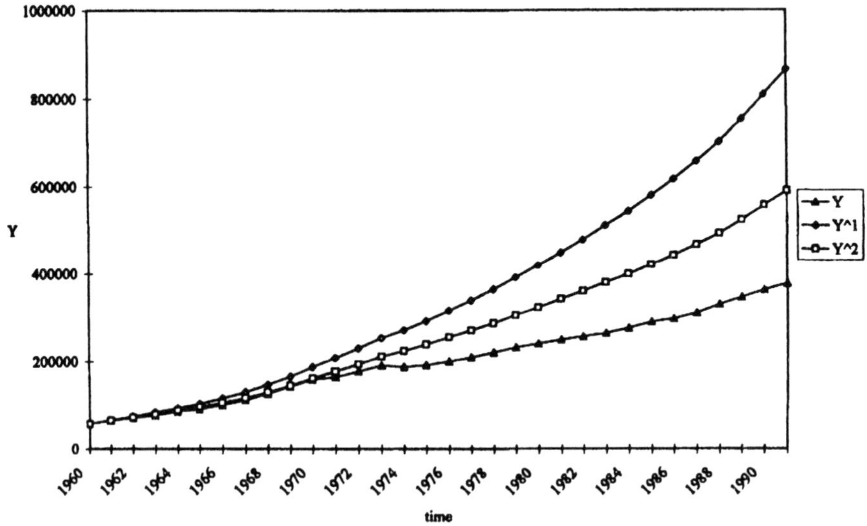

FIGURE 9. Actual vs. Estimated Y (Sigma = 0.8).

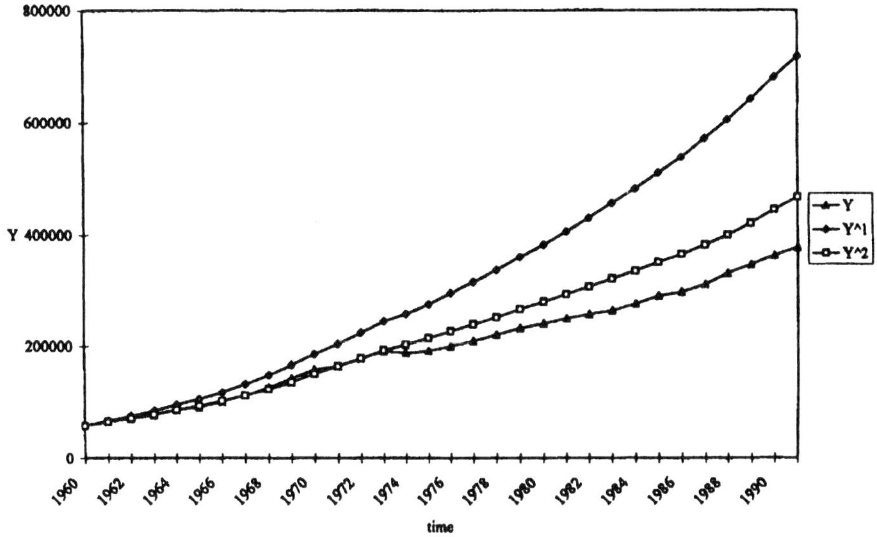

FIGURE 10. Actual vs. Estimated Y (Sigma = 0.6).

in productivity growth for $\sigma = 0.5$. These diagrams show that the estimates of Y assuming biased technical progress is close to actual when $\sigma = 0.902$ but deteriorates for lower values of σ. Assumption of Hicks neutral technical progress does not give a good fit. However, unlike the case of United States, the growth rates of efficiencies for various values of elasticity shows considerable variation, showing that the estimates are not robust.

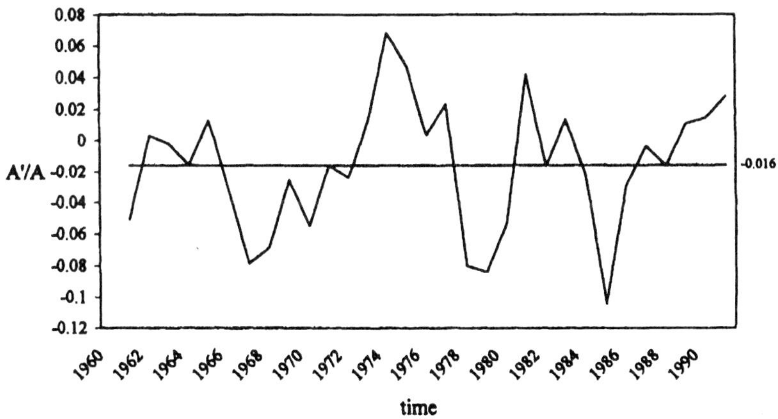

FIGURE 11A. Growth Rate of Capital Augmenting Technical Progress.

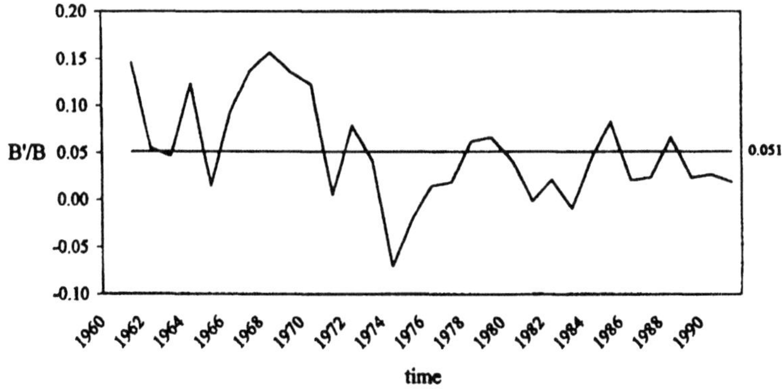

FIGURE 11B. Growth Rate of Labor Augmenting Technical Progress.

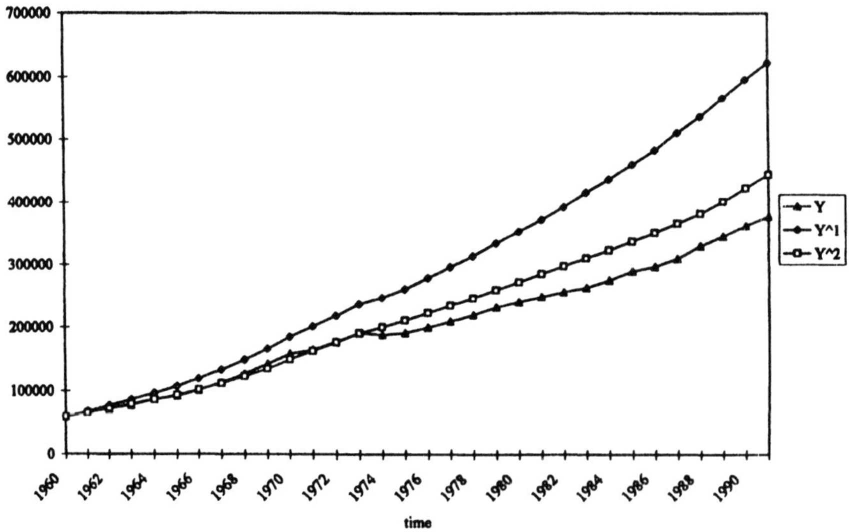

FIGURE 12. Actual vs. Estimated Y (Sigma = 0.5).

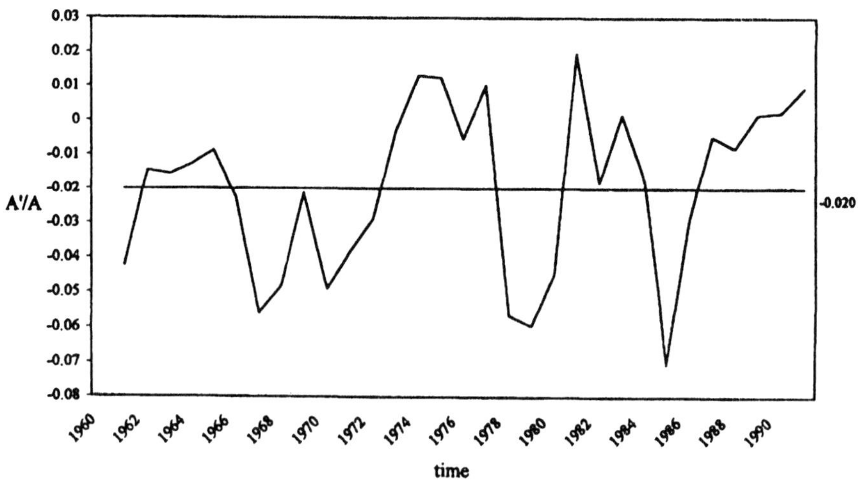

FIGURE 13A. Growth Rate of Capital Augmenting Technical Progress.

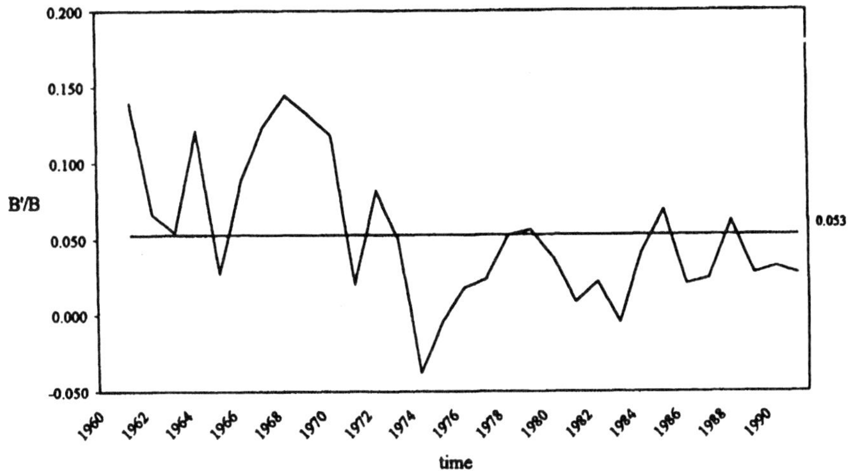

FIGURE 13B. Growth Rate of Labor Augmenting Technical Progress.

5. Conclusion

The historical evidence and our econometric estimate suggest that the rate of growth of A, the efficiency index of capital, is not constant. It fluctuates much more than that of labor and shows definite trends in different periods of time. So there is no apriori justification in assuming that all technical progress is Harrod neutral.

However the assumption of biased technical progress raises both the oretical and empirical problems. As is well known, the Solow-Swan model has steady state only if technical progress is labor augmenting. The question arises whether it is necessary to assume biased technical progress if it faces both econometric and theoretical problems. In two theoretical papers [Sato and Ramachandran (1998) and Sato, Ramachandran and Lian (1988)], we argue that neoclassical models with steady state can be constructed. In this paper, we examine the empirical evidence. The technique we used was developed in Sato (1980). For United States it provides fairly robust estimates but for Japan the results seem very sensitive to the value of sigma used. As an ongoing research program, we are now estimating growth rates of factor productivities using this method for other OECD countries.

References

[1] Ahmed, S (1966) "On the Theory Induced Innovation," *Economic Journal*, **76**, 344–57.

[2] Drandakis E.M. and E.S. Phelps, (1966) "A Model of Induced Invention, Growth and Distribution," *Economic Journal*, **76**, 823–40.

[3] Habakkuk, H.J. (1962) *The American and British Technology in the Nineteenth Century*. Cambridge: Cambridge University Press.

[4] Hicks, J.R. (1932) *Theory of Wages*. London: McMillian and Co.

[5] Jone, H.G. (1976) *An Introduction to Modern Theories of Economic Growth*, New York: McGraw-Hill Book Company.
[6] Kamine, M.I and N.L. Schwartz (1969) "Induced Factor Augmenting Technological Progress from a Microeconomic Viewpoint," *Econometrica*, **37**, 668–84.
[7] Kennedy, C. (1960) "Induced bias in Innovation and And the Theory of Distribution," *Economic Journal*, **74**, 541–7
[8] Lucus, Jn., R.E. (1993) "Making a Miracle," *Econometrica*, **61**, 251–72.
[9] Mantoux, P. (1961) *The Industrial Revolution in the Eighteenth Century*. New York: Harper and Row.
[10] MacLeod, C (1988) *Inventing the Industrial Revolution*, Cambridge: Cambridge University Press
[11] Mokyar, J (1990) *The Lever of Riches*. New York: Oxford University Press.
[12] Nordhaus, W.D. (1969) *Invention, Growth and Welfare: A Theoretical Treatment of Technological Change*, Cambridge, MA: MIT Press.
[13] Pilat, D (1994) *The Economics of Rapid Growth: The Experience of Japan and Korea*, Edward Elgar.
[14] Romer, P (1986) "Increasing Returns and Long-Run Growth," *Journal of Political Economy*, **94**, 1002–37.
[15] Rothbarth, E. (1946) "Causes of the Superior Efficiency of U.S.S Industry as compared with British Industry," *Economic Journal*, **56**, 383–90.
[16] Rutten, V. W. (1997) "Induced Innovation, Evalutionary Theory and Path Dependence: Sources of Technical Change," *Economic Journal*, **107**, 1520–29.
[17] Samuelson, P.S. (1965) "A Theory of Induced Innovation along Kennedy-Weisäker lines," *Review of Economics and Statistics*, 345–56.
[18] Sato, R (1970) "The estimation of biased technical progress and the production function," *Internatinal Economic Review*, **11**, 179–208. Reprinted in *Growth Theory and Technical Change: The Selected Essays of Ryuzo Sato* (Economics of the Twentieth Century), Cheltenham, UK: Edward Elgar, 1996, Chapter 12.
[19] Sato, R., and R.V. Ramachandran (1998) "Optimal Growth with Endogenous Technical Progress: Hicksian Bias in a Macro-Model," *The Japanese Economic Review*, Forthcoming 1999.
[20] Sato, R., R.V. Ramachandran, and C. Lian (1998)) "A Model of Optimal Economic Growth with Endogenous Bias," *Macroeconomic Dynamics*, Forthcoming 1999.
[21] Salter, W.E.G. (1960) *Productivity and Technical Change*. Cambridge: Cambridge University Press.
[22] Sato, R. and R. Ramachandran (1987) "Factor Price Variation and the Hicksian Hypothesis: A Microeconomic Model," *Oxford Economic Papers*, **39**, 343–356.
[23] Weizäcker, C.C. (1966) "A Tentative Note on a Two-Sector Model of Induced Technical Progress," *Review of Economics and Statistics*, 245–51.
[24] Wright, G., (1997) "Towards a More Historical Approach to Technological Change," *Economic Journal*, **107**, 1560–65.

Table A1. United States 1909–1993.

	Y	K	L	r	w
1909	44664	138436	66563	0.108	0.446
1910	44910	145520	68831	0.102	0.437
1911	47195	146167	70102	0.108	0.448
1912	48226	153595	72786	0.104	0.444
1913	51937	156659	73839	0.111	0.468
1914	45886	147066	71210	0.101	0.435
1915	46676	151603	70859	0.106	0.432
1916	56293	171325	78007	0.118	0.463
1917	52695	175687	79459	0.111	0.418
1918	56365	180011	78283	0.107	0.474
1919	58985	180546	75422	0.116	0.505
1920	59734	184298	76336	0.103	0.533
1921	58698	155887	68167	0.139	0.543
1922	62084	168694	74269	0.125	0.553
1923	71285	197128	81994	0.122	0.576
1924	74284	197560	79197	0.124	0.628
1925	75481	211265	82429	0.120	0.608
1926	81366	226996	86127	0.117	0.636
1927	81833	227826	86508	0.116	0.640
1928	83097	235339	87083	0.119	0.632
1929	88562	249382	89467	0.118	0.661
1930	79817	242728	81854	0.114	0.637
1931	73021	215616	72386	0.110	0.681
1932	60665	183450	62069	0.131	0.589
1933	57772	174308	61248	0.120	0.602
1934	65041	181611	62366	0.127	0.673
1935	74221	184713	66023	0.141	0.730
1936	83278	195770	73426	0.152	0.729
1937	90884	207780	77568	0.149	0.773
1938	83743	192708	70460	0.144	0.795
1939	91530	199232	75131	0.159	0.796
1940	101313	210558	79694	0.172	0.817
1941	116415	232213	89276	0.189	0.812
1942	127434	254535	97056	0.178	0.846
1943	136274	263965	101633	0.177	0.882
1944	146470	263740	100124	0.184	0.977
1945	145052	255976	94920	0.178	1.048
1946	140288	255196	96671	0.172	0.998
1947	142022	266942	100072	0.174	0.955
1948	149895	279048	101304	0.178	0.988
1949	147122	281433	96784	0.170	1.025
1950	163260	294869	100352	0.201	1.036
1951	173398	317629	104801	0.188	1.084
1952	178864	332480	106168	0.171	1.151
1953	186264	343207	109195	0.169	1.175
1954	184482	344371	103523	0.163	1.239
1955	200993	362673	107954	0.182	1.249

(*Continued*)

TABLE A1. (*Cont'd*)

	Y	K	L	r	w
1956	205730	376986	124341	0.174	1.127
1957	210125	389533	111104	0.168	1.303
1958	205700	383892	106250	0.161	1.353
1959	221385	402047	110732	0.174	1.368
1960	227593	412304	111881	0.171	1.406
1961	232781	422029	111321	0.172	1.439
1962	244742	433173	112895	0.179	1.483
1963	254528	445207	114595	0.181	1.517
1964	268590	460172	116586	0.187	1.567
1965	283748	479868	119414	0.191	1.609
1966	299750	502268	120956	0.192	1.683
1967	305647	523687	120757	0.183	1.736
1968	317281	546269	122194	0.179	1.797
1969	325141	570888	124923	0.167	1.838
1970	332731	594391	123748	0.154	1.867
1971	332513	616794	123745	0.158	1.897
1972	348716	640623	128424	0.160	1.917
1973	368183	670614	132768	0.159	1.972
1974	365547	700250	133744	0.148	1.960
1975	361493	723123	129571	0.149	1.959
1976	380661	746063	134552	0.154	1.975
1977	399425	772993	139638	0.158	1.985
1978	420528	805885	145155	0.161	2.005
1979	432363	842147	149257	0.157	2.010
1980	429181	875970	147913	0.147	2.028
1981	437721	910397	149819	0.147	2.028
1982	426024	938605	146930	0.136	2.033
1983	443328	963751	149998	0.144	2.028
1984	472895	996578	158005	0.151	2.041
1985	487555	1033296	159792	0.150	2.081
1986	501516	1065114	163107	0.150	2.097
1987	516877	1094409	167534	0.149	2.113
1988	537822	1125157	170832	0.150	2.160
1989	551377	1154665	173764	0.152	2.161
1990	556657	1183017	173372	0.147	2.206
1991	550689	1208491	170387	0.142	2.227
1992	565538	1235506	171705	0.143	2.266
1993	582040	1268760	174807	0.144	2.284

Source: OECD.

TABLE A2. Estimates of Sigma, Growth Rates of Factor Augmenting Technical Progress, and Output.

	σ1	σ2	A'/A	B'/B	Y	Y^1	Y^2
1909					44664	44664	44664
1910	0.457	0.457	−0.029	−0.035	44910	47105	47120
1911	0.369	0.369	0.030	0.040	47195	48198	48221

TABLE A2. (*Cont'd*)

	σ1	σ2	A′/A	B′/B	Y	Y^1	Y^2
1912	0.372	0.372	−0.013	−0.023	48226	50927	50966
1913	−0.408	0.000	0.043	0.068	51937	52403	52455
1914	−2.014	0.000	−0.033	−0.096	45886	50442	50471
1915	−0.702	0.000	−0.071	0.051	46676	51676	51745
1916	−0.759	0.000	0.023	0.119	56293	58292	58411
1917	−0.164	0.000	−0.118	−0.064	52695	60262	60401
1918	0.233	0.233	0.123	0.037	56365	61200	61425
1919	−2.901	0.000	0.007	0.106	58985	60786	61109
1920	0.054	0.054	0.091	−0.054	59734	62483	62851
1921	0.145	0.145	−0.022	0.182	58698	55024	55223
1922	−0.062	0.000	0.057	−0.076	62084	60449	60643
1923	0.972	0.972	−0.012	0.037	71285	69343	69763
1924	0.504	0.504	0.062	0.067	74284	68913	69476
1925	509.264	2.000	−0.067	−0.015	75481	73451	74178
1926	0.427	0.427	0.030	0.018	81366	78608	79538
1927	−0.044	0.000	0.014	−0.005	81833	79845	80800
1928	−0.623	0.000	−0.063	0.031	83097	82234	83381
1929	0.549	0.549	0.024	0.028	88562	86615	88041
1930	−12.071	0.000	−0.116	0.007	79817	82277	84058
1931	0.038	0.038	0.096	−0.001	73021	73741	75394
1932	0.020	0.020	−0.241	0.073	60665	63759	65168
1933	−0.341	0.000	0.091	−0.091	57772	62647	63831
1934	0.407	0.407	0.102	0.093	65041	65156	66535
1935	1.681	1.681	0.135	0.071	74221	68607	69792
1936	0.677	0.677	0.040	0.018	83278	75594	76516
1937	0.061	0.061	0.078	0.006	90884	80950	81949
1938	0.313	0.313	0.020	0.000	83743	75062	76130
1939	0.301	0.301	0.006	0.050	91530	79883	80708
1940	0.078	0.078	0.017	0.060	101313	85579	86379
1941	0.163	0.163	−0.017	0.058	116415	96311	96970
1942	0.092	0.092	0.055	−0.027	127434	106307	107065
1943	−0.192	0.000	0.072	−0.001	136274	112128	112702
1944	0.221	0.221	0.107	0.074	146470	112446	113136
1945	0.209	0.209	0.076	0.016	145052	108956	109879
1946	1.865	1.865	−0.024	−0.053	140288	111208	111728
1947	−0.188	0.000	−0.079	−0.001	142022	116992	117600
1948	3.382	2.000	−0.006	0.050	149895	121490	122578
1949	0.657	0.657	−0.009	0.018	147122	120145	122098
1950	−0.065	0.000	−0.062	0.129	163260	126569	128733
1951	0.302	0.302	0.035	−0.012	173398	135428	138308
1952	0.216	0.216	0.066	−0.026	178864	140633	144277
1953	0.118	0.118	0.028	0.004	186264	146534	150322
1954	0.645	0.645	0.006	0.036	184482	143697	148717
1955	−0.096	0.000	−0.047	0.081	200993	152162	157658
1956	2.125	2.000	0.015	−0.125	205730	170245	173149
1957	0.725	0.725	0.013	0.130	210125	162866	169836
1958	0.379	0.379	0.025	0.009	205700	159320	166963

(*Continued*)

Table A2. Estimates of Sigma, Growth Rates of Factor Augmenting Technical Progress, and Output. (*Cont'd*)

	σ1	σ2	A′/A	B′/B	Y	Y^1	Y^2
1959	−0.075	0.000	−0.024	0.055	221385	168268	176417
1960	0.318	0.318	0.025	0.007	227593	172919	181711
1961	1.965	1.965	−0.011	0.032	232781	175827	185681
1962	−1.812	0.000	0.011	0.043	244742	181151	191667
1963	1.594	1.594	0.009	0.026	254528	186828	198071
1964	7.813	2.000	0.011	0.042	268590	193342	205536
1965	4.319	2.000	0.004	0.036	283748	201570	214974
1966	0.786	0.786	0.016	0.040	299750	208848	224163
1967	0.589	0.589	−0.001	0.011	305647	213967	231687
1968	0.532	0.532	0.014	0.017	317281	221179	240991
1969	0.262	0.262	0.025	−0.018	325141	230340	252116
1970	0.531	0.531	−0.014	−0.012	322731	234396	259327
1971	−3.168	0.000	−0.042	0.045	332513	239754	267424
1972	1.028	1.028	0.010	0.011	348716	251766	280871
1973	0.351	0.351	0.026	0.014	368183	264272	295649
1974	0.586	0.586	−0.029	−0.023	365547	272116	306947
1975	−7.109	0.000	−0.093	0.042	361493	271481	310822
1976	0.258	0.258	0.007	0.020	380661	284683	325503
1977	0.082	0.082	−0.001	0.017	399425	298738	341477
1978	−0.463	0.000	0.003	0.016	420528	314396	359649
1979	0.662	0.662	−0.010	−0.003	432363	328473	377250
1980	0.701	0.701	−0.030	−0.006	429181	333561	387732
1981	12.646	2.000	−0.035	0.014	437721	344065	402578
1982	0.628	0.628	−0.034	−0.018	426024	345668	409643
1983	−0.088	0.000	−0.038	0.041	443328	357489	424401
1984	0.509	0.509	0.018	0.019	472895	379199	448151
1985	1.017	1.017	−0.006	0.019	487555	390335	464395
1986	1.023	1.023	−0.002	0.008	501516	404020	482007
1987	0.025	0.025	0.013	−0.001	516877	419837	501003
1988	0.644	0.644	0.015	0.019	537822	433932	519029
1989	−0.631	0.000	−0.017	0.015	551377	447455	536547
1990	0.497	0.497	0.004	0.003	556657	454479	549011
1991	0.819	0.819	−0.025	0.004	550689	455903	556741
1992	1.482	1.482	0.001	0.021	565538	466282	571823
1993	−13.582	0.000	−0.004	0.014	582040	481151	591609
Average		0.502	0.004	0.019			

Source: OECD.

Note:

1. σ1 is calculated using equation (8);
2. σ2 is a value of σ1 bounded above by two and below by zero;
3. A′/A and B′/B are calculated using equations (6) and (7);
4. Y^1 is an estimate of Y calculated setting an initial value for the year 1909 and assuming Hicks neutral technical progress;
5. Y^2 is an estimate of Y calculated using biased technical progress as estimated in this table.

TABLE A3. Estimates of Growth Rates of Factor Augmenting Technical Progress and Output Assuming $\sigma = 0.4$.

	$\sigma = 0.4$				
	A′/A	B′/B	Y	Y^1	Y^2
1909			44664	44664	44664
1910	−0.034	−0.032	44910	47033	47125
1911	0.036	0.037	47195	48187	48363
1912	−0.018	−0.021	48226	50859	51143
1913	0.047	0.066	51937	52306	52694
1914	−0.042	−0.092	45886	50474	50893
1915	−0.052	0.041	46676	51540	52101
1916	0.038	0.111	56293	57992	58772
1917	−0.108	−0.070	52695	59912	60831
1918	0.097	0.054	56365	60606	61749
1919	0.019	0.099	58985	59943	61308
1920	0.057	−0.036	59734	61560	63108
1921	0.040	0.154	58698	54520	55792
1922	0.030	−0.060	62084	59937	61389
1923	−0.014	0.038	71285	68349	70399
1924	0.054	0.071	74284	67654	69995
1925	−0.061	−0.018	75481	71890	74671
1926	0.021	0.022	81366	76692	79994
1927	0.010	−0.003	81833	77906	81396
1928	−0.047	0.024	83097	79996	83936
1929	0.018	0.031	88562	83949	88521
1930	−0.102	0.000	79817	79144	84167
1931	0.073	0.011	73021	70893	75586
1932	−0.168	0.038	60665	61356	65489
1933	0.061	−0.072	57772	60567	64464
1934	0.094	0.098	65041	62816	67164
1935	0.130	0.073	74221	66466	70795
1936	0.047	0.015	83278	73653	78010
1937	0.061	0.015	90884	78829	83631
1938	0.011	0.005	83743	72916	77680
1939	0.023	0.041	91530	77881	82631
1940	0.027	0.054	101313	83470	88566
1941	0.003	0.047	116415	94107	99636
1942	0.036	−0.016	127434	103776	110079
1943	0.058	0.006	136274	109580	116067
1944	0.097	0.080	146470	109713	116547
1945	0.057	0.026	145052	106017	113165
1946	−0.026	−0.052	140288	108478	115322
1947	−0.063	−0.008	142022	113982	121441
1948	−0.001	0.048	149895	117918	126499
1949	−0.015	0.021	147122	115850	125757
1950	−0.021	0.109	163260	121887	132658
1951	0.019	−0.002	173398	129916	142425
1952	0.039	−0.011	178864	134350	148446
1953	0.022	0.007	186264	139922	154810

(Continued)

Table A3. (*Cont'd*)

	$\sigma = 0.4$				
	A′/A	B′/B	Y	Y^1	Y^2
1954	0.000	0.039	184482	136207	152789
1955	−0.020	0.069	200993	144045	162068
1956	0.005	−0.120	205730	163307	179249
1957	0.005	0.134	210125	153308	174523
1958	0.015	0.014	205700	149366	171435
1959	−0.006	0.047	221385	157651	181311
1960	0.017	0.011	227593	161684	186798
1961	−0.007	0.031	232781	163771	190750
1962	0.016	0.041	244742	168454	196986
1963	0.010	0.026	254528	173437	203652
1964	0.014	0.040	268590	179095	211366
1965	0.007	0.034	283748	186258	221080
1966	0.013	0.041	299750	192111	230296
1967	−0.008	0.014	305647	195640	237588
1968	0.008	0.020	317281	201383	246900
1969	0.010	−0.011	325141	209083	258219
1970	−0.025	−0.007	322731	211296	264930
1971	−0.030	0.040	332513	215018	272765
1972	0.010	0.011	348716	225765	286892
1973	0.020	0.016	368183	236569	302122
1974	−0.036	−0.020	365547	242382	313191
1975	−0.076	0.035	361493	239700	315858
1976	0.012	0.018	380661	251580	331495
1977	0.004	0.015	399425	264060	348353
1978	0.005	0.015	420528	277789	367359
1979	−0.012	−0.002	432363	289585	385392
1980	−0.035	−0.003	429181	292136	395043
1981	−0.030	0.012	437721	300288	409872
1982	−0.042	−0.014	426024	299670	415818
1983	−0.020	0.034	443328	309679	431276
1984	0.023	0.017	472895	329298	456917
1985	−0.006	0.019	487555	337843	473168
1986	−0.002	0.008	501516	349234	491461
1987	0.009	0.000	516877	362889	511620
1988	0.014	0.019	537822	374664	530471
1989	−0.012	0.013	551377	385887	548800
1990	−0.002	0.006	556657	390580	561118
1991	−0.027	0.005	550689	389835	567951
1992	0.002	0.020	565538	397970	583507
1993	−0.002	0.013	582040	410202	604189
Aver.	0.004	0.020			

Source: OECD.

Note:

1. A′/A and B′/B are calculated using equations (6) and (7);
2. Y^1 is an estimate of Y calculated setting an initial value for the year 1909 and assuming Hicks neutral technical progress;
3. Y^2 is an estimate of Y calculated using biased technical progress as estimated in this table.

TABLE A4. Estimates of Growth Rates of Factor Augmenting Technical Progress and Output Assuming $\sigma = 0.6$.

	$\sigma = 0.6$				
	A′/A	B′/B	Y	Y^1	Y^2
1909			44664	44664	44664
1910	−0.022	−0.039	44910	47152	47108
1911	0.022	0.043	47195	48205	48111
1912	−0.006	−0.027	48226	50972	50824
1913	0.037	0.071	51937	52466	52262
1914	−0.021	−0.102	45886	50421	50154
1915	−0.100	0.065	46676	51765	51457
1916	0.002	0.131	56293	58489	58102
1917	−0.133	−0.055	52695	60491	60034
1918	0.162	0.013	56365	61593	61113
1919	−0.012	0.116	58985	61348	60866
1920	0.139	−0.081	59734	63099	62556
1921	−0.111	0.222	58698	55357	54735
1922	0.097	−0.098	62084	60787	60020
1923	−0.009	0.035	71285	70005	69174
1924	0.072	0.062	74284	69759	68948
1925	−0.076	−0.011	75481	74505	73639
1926	0.044	0.011	81366	79909	78991
1927	0.020	−0.008	81833	81161	80144
1928	−0.085	0.042	83097	83762	82727
1929	0.033	0.023	88562	88449	87402
1930	−0.137	0.018	79817	84466	83657
1931	0.128	−0.018	73021	75732	74960
1932	−0.348	0.124	60665	65436	64680
1933	0.135	−0.119	57772	64086	63139
1934	0.112	0.087	65041	66783	65820
1935	0.141	0.068	74221	70082	68815
1936	0.032	0.023	83278	76916	75187
1937	0.102	−0.008	90884	82396	80460
1938	0.033	−0.006	83743	76531	74741
1939	−0.019	0.062	91530	81244	79050
1940	0.002	0.067	101313	87012	84511
1941	−0.046	0.074	116415	97804	94727
1942	0.082	−0.044	127434	108024	104526
1943	0.092	−0.012	136274	113853	109895
1944	0.123	0.066	146470	114302	110276
1945	0.103	0.002	145052	110962	107098
1946	−0.021	−0.055	140288	113063	108727
1947	−0.102	0.010	142022	119042	114387
1948	−0.014	0.054	149895	123941	119254
1949	0.000	0.013	147122	123130	118910
1950	−0.121	0.158	163260	129831	125309
1951	0.059	−0.026	173398	139300	134663
1952	0.105	−0.047	178864	145084	140523
1953	0.037	−0.001	186264	151222	146295

(*Continued*)

TABLE A4. (*Cont'd*)

	$\sigma = 0.6$				
	A′/A	B′/B	Y	Y^1	Y^2
1954	0.016	0.031	184482	149088	144927
1955	−0.088	0.099	200993	158021	153555
1956	0.030	−0.131	205730	175117	167864
1957	0.026	0.123	210125	169848	165391
1958	0.041	0.001	205700	166654	162642
1959	−0.049	0.066	221385	176102	171718
1960	0.036	0.002	227593	181245	176809
1961	−0.015	0.035	232781	184835	180715
1962	0.005	0.046	244742	190673	186452
1963	0.007	0.027	254528	196907	192595
1964	0.006	0.044	268590	204117	199793
1965	−0.001	0.038	283748	213216	208923
1966	0.019	0.038	299750	221706	217951
1967	0.010	0.006	305647	228240	225487
1968	0.024	0.012	317281	236747	234637
1969	0.047	−0.028	325141	247175	245469
1970	0.002	−0.019	322731	252990	252853
1971	−0.060	0.052	332513	259912	260965
1972	0.010	0.011	348716	272959	273785
1973	0.034	0.010	368183	286946	288053
1974	−0.020	−0.027	365547	296748	299294
1975	−0.119	0.053	361493	298385	303811
1976	0.000	0.023	380661	312644	317656
1977	−0.007	0.020	399425	328012	332818
1978	0.000	0.017	420528	345329	350175
1979	−0.007	−0.004	432363	361515	367211
1980	−0.022	−0.009	429181	369349	377992
1981	−0.043	0.018	437721	382224	392576
1982	−0.024	−0.023	426024	386465	400166
1983	−0.063	0.052	443328	399978	414207
1984	0.012	0.021	472895	423261	436371
1985	−0.006	0.019	487555	437089	452293
1986	−0.002	0.008	501516	452985	469142
1987	0.017	−0.003	516877	470741	487054
1988	0.017	0.018	537822	487061	504211
1989	−0.025	0.019	551377	502817	520872
1990	0.013	−0.002	556657	512472	533134
1991	−0.022	0.002	550689	516682	541195
1992	0.000	0.021	565538	529445	555645
1993	−0.007	0.015	582040	546949	574461
Aver.	0.004	0.019			

Source: OECD.

Note:

1. A′/A and B′/B are calculated using equations (6) and (7);
2. Y^1 is an estimate of Y calculated setting an initial value for the year 1909 and assuming Hicks neutral technical progress;
3. Y^2 is an estimate of Y calculated using biased technical progress as estimated in this table.

TABLE A5. Estimates of Growth Rates of Factor Augmenting Technical Progress and Output Assuming $\sigma = 0.8$.

	$\sigma = 0.8$				
	A′/A	B′/B	Y	Y^1	Y^2
1909			44664	44664	44664
1910	0.014	−0.057	44910	47209	47032
1911	−0.017	0.063	47195	48214	47851
1912	0.030	−0.045	48226	51027	50451
1913	0.005	0.087	51937	52544	51755
1914	0.043	−0.133	45886	50395	49444
1915	−0.244	0.137	46676	51875	50709
1916	−0.107	0.189	56293	58732	57202
1917	−0.210	−0.012	52695	60775	58969
1918	0.360	−0.107	56365	62082	60025
1919	−0.103	0.165	58985	62049	59786
1920	0.384	−0.216	59734	63868	61312
1921	−0.567	0.424	58698	55770	53303
1922	0.295	−0.214	62084	61206	58269
1923	0.006	0.027	71285	70832	67220
1924	0.126	0.033	74284	70820	66984
1925	−0.119	0.011	75481	75830	71478
1926	0.111	−0.024	81366	81552	76611
1927	0.051	−0.022	81833	82824	77510
1928	−0.200	0.098	83097	85699	79931
1929	0.077	0.000	88562	90783	84399
1930	−0.241	0.074	79817	87276	80927
1931	0.291	−0.105	73021	78291	72330
1932	−0.889	0.382	60665	67588	62197
1933	0.356	−0.259	57772	65921	60384
1934	0.164	0.057	65041	68867	62874
1935	0.173	0.051	74221	71959	65384
1936	−0.014	0.046	83278	78587	71046
1937	0.224	−0.076	90884	84226	75853
1938	0.099	−0.041	83743	78397	70364
1939	−0.147	0.123	91530	82965	74109
1940	−0.073	0.107	101313	88823	79018
1941	−0.192	0.153	116415	99686	88289
1942	0.221	−0.129	127434	110192	97224
1943	0.193	−0.068	136274	116027	101931
1944	0.202	0.025	146470	116646	102105
1945	0.242	−0.068	145052	113503	99028
1946	−0.005	−0.061	140288	115406	100209
1947	−0.218	0.063	142022	121634	105225
1948	−0.052	0.074	149895	127055	109590
1949	0.044	−0.010	147122	126952	109271
1950	−0.422	0.305	163260	134014	114929
1951	0.182	−0.098	173398	144288	123386
1952	0.305	−0.156	178864	150848	128642

(Continued)

TABLE A5. (*Cont'd*)

	$\sigma = 0.8$				
	A′/A	B′/B	Y	Y^1	Y^2
1953	0.085	−0.023	186264	157297	133624
1954	0.063	0.008	184482	156141	132417
1955	−0.291	0.189	200993	165700	140023
1956	0.104	−0.164	205730	181395	152005
1957	0.088	0.089	210125	179069	150395
1958	0.121	−0.036	205700	176395	147760
1959	−0.177	0.121	221385	186515	155653
1960	0.091	−0.024	227593	192344	159991
1961	−0.040	0.046	232781	196912	163361
1962	−0.028	0.061	244742	203469	168230
1963	−0.001	0.031	254528	210487	173450
1964	−0.018	0.055	268590	218681	179631
1965	−0.025	0.050	283748	229014	187546
1966	0.034	0.031	299750	239270	195504
1967	0.063	−0.021	305647	247924	202243
1968	0.071	−0.010	317281	258364	210280
1969	0.159	−0.079	325141	270669	219702
1970	0.083	−0.055	322731	279247	226414
1971	−0.147	0.088	332513	288633	233614
1972	0.010	0.011	348716	303161	244439
1973	0.077	−0.008	368183	319359	256692
1974	0.030	−0.048	365547	332278	266620
1975	−0.245	0.106	361493	337842	271101
1976	−0.034	0.037	380661	353580	282534
1977	−0.041	0.034	399425	370853	295170
1978	−0.017	0.024	420528	390633	309776
1979	0.009	−0.012	432363	410115	324311
1980	0.017	−0.027	429181	422681	334028
1981	−0.084	0.036	437721	439493	346583
1982	0.030	−0.048	426024	448558	353586
1983	−0.190	0.107	443328	464753	365110
1984	−0.021	0.036	472895	490082	383048
1985	−0.006	0.019	487555	508498	396623
1986	−0.002	0.008	501516	527982	410537
1987	0.041	−0.014	516877	548714	425037
1988	0.025	0.014	537822	568636	439045
1989	−0.065	0.037	551377	588040	452580
1990	0.060	−0.024	556657	602439	462806
1991	−0.007	−0.005	550689	612059	469783
1992	−0.008	0.025	565538	628993	481475
1993	−0.023	0.023	582040	650923	496696
Aver. 0.004	0.020				

Source: OECD.

Note:

1. A′/A and B′/B are calculated using equations (6) and (7);
2. Y^1 is an estimate of Y calculated setting an initial value for the year 1909 and assuming Hicks neutral technical progress;
3. Y^2 is an estimate of Y calculated using biased technical progress as estimated in this table.

Table A6. Japan 1960–1991.

	Y	K	L	r	w
1960	58275	57199	99905	0.367	0.373
1961	65530	65978	99561	0.364	0.417
1962	71146	75371	99185	0.333	0.464
1963	76940	85158	100155	0.310	0.505
1964	86186	96105	100350	0.309	0.563
1965	90982	107051	100657	0.279	0.607
1966	100860	119243	103378	0.283	0.649
1967	112498	134387	104979	0.293	0.697
1968	127261	153202	106069	0.303	0.763
1969	143298	174853	105956	0.301	0.856
1970	158894	201425	105821	0.293	0.943
1971	164740	227599	104521	0.256	1.019
1972	177898	255685	103721	0.244	1.115
1973	191169	284272	104295	0.228	1.212
1974	188292	310334	100065	0.180	1.322
1975	191715	335380	99086	0.157	1.402
1976	200196	358296	100979	0.151	1.447
1977	209695	381245	102356	0.145	1.509
1978	220307	404148	103886	0.150	1.536
1979	232411	430788	105718	0.156	1.562
1980	240549	458278	106241	0.155	1.597
1981	248868	486521	106923	0.144	1.673
1982	256780	513125	107848	0.140	1.714
1983	263619	538621	110219	0.134	1.738
1984	275283	566646	110941	0.134	1.796
1985	289714	598159	111998	0.143	1.824
1986	296989	630719	112608	0.139	1.859
1987	309918	662517	114688	0.138	1.906
1988	330107	699894	115755	0.141	1.998
1989	345852	745419	117162	0.136	2.084
1990	362567	799118	117815	0.130	2.196
1991	377220	855466	117537	0.121	2.325

Source: OECD.

Table A7. Estimates of Sigma, Growth Rates of Factor Augmenting Technical Progress, and Output.

	σ1	σ2	A′/A	B′/B	Y	Y^1	Y^2
1960				58275	58275	58275	
1961	1.252	1.252	−0.182	0.230	65530	66196	66354
1962	0.740	0.740	0.272	−0.123	71146	74646	74994
1963	0.762	0.762	0.207	−0.082	76940	83853	84426
1964	1.084	1.084	−0.057	0.147	86186	93864	94700
1965	0.638	0.638	0.347	−0.181	90982	104136	105272
1966	1.536	1.536	−0.171	0.167	100860	116372	117869
1967	2.968	2.000	−0.430	0.332	112498	130593	132509

(*Continued*)

TABLE A7. (*Cont'd*)

	σ1	σ2	A′/A	B′/B	Y	Y^1	Y^2
1968	2.084	2.000	−0.380	0.346	127261	147418	149826
1969	1.117	1.117	−0.086	0.175	143298	165866	168823
1970	1.195	1.195	−0.144	0.182	158894	187696	191291
1971	0.689	0.689	0.324	−0.225	164740	208806	213062
1972	0.916	0.916	0.053	0.031	177898	231658	236646
1973	0.705	0.705	0.247	−0.098	191169	256298	262116
1974	0.441	0.441	0.920	−0.567	188292	275654	282155
1975	0.479	0.479	0.588	−0.270	191715	297602	304932
1976	0.690	0.690	0.136	−0.038	200196	321999	330331
1977	0.594	0.594	0.222	−0.059	209695	346818	356205
1978	−2.139	0.000	−0.445	0.198	220307	372782	383310
1979	−2.206	0.000	−0.462	0.217	232411	402479	414290
1980	1.846	1.846	−0.188	0.099	240549	432067	445158
1981	0.467	0.467	0.388	−0.154	248868	463411	477881
1982	0.928	0.928	0.011	0.010	256780	495268	511207
1983	0.465	0.465	0.195	−0.082	263619	530111	547755
1984	1.478	1.478	−0.100	0.074	275283	565171	584474
1985	−0.943	0.000	−0.623	0.290	289714	604476	625613
1986	1.064	1.064	−0.036	0.023	296989	645087	668114
1987	0.941	0.941	0.011	0.017	309918	689856	715093
1988	1.996	1.996	−0.142	0.121	330107	738035	765516
1989	0.684	0.684	0.148	−0.039	345852	794469	824470
1990	0.665	0.665	0.202	−0.057	362567	856736	889329
1991	0.585	0.585	0.316	−0.105	377220	920309	955425
Average	0.902	0.0368	0.019				

Source: OECD.

Note:

1. s1 is calculated using equation (8);
2. s2 is a value of s1 bounded above by two and below by zero;
3. A′/A and B′/B are calculated using equations (6) and (7);
4. Y^1 is an estimate of Y calculated setting an initial value for the year 1909 and assuming Hicks neutral technical progress;
5. Y^2 is an estimate of Y calculated using biased technical progress as estimated in this table.

TABLE **A8.** Estimates of Growth Rates of Factor Augmenting Technical Progress and Output Assuming $\sigma = 0.8$.

	$\sigma = 0.8$				
	A′/A	B′/B	Y	Y^1	Y^2
1960			58275	58275	58275
1961	−0.093	0.173	65530	66334	65258
1962	0.090	−0.003	71146	74906	72531
1963	0.066	0.005	76940	84207	80280
1964	−0.029	0.131	86186	94295	88551
1965	0.122	−0.049	90982	104611	96794
1966	−0.077	0.118	100860	116873	106557

Table A8. (*Cont'd*)

	$\sigma = 0.8$				
	A'/A	B'/B	Y	Y^1	Y^2
1967	−0.193	0.200	112498	131073	117825
1968	−0.170	0.218	127261	147790	131073
1969	−0.045	0.148	143298	165984	145338
1970	−0.084	0.141	158894	187347	162076
1971	0.095	−0.070	164740	207785	177667
1972	0.001	0.063	177898	229758	194219
1973	0.089	−0.004	191169	253414	211736
1974	0.345	−0.232	188292	271339	224376
1975	0.223	−0.102	191715	291966	238729
1976	0.046	−0.003	200196	315296	254625
1977	0.088	−0.007	209695	338903	270363
1978	−0.199	0.106	220307	363581	286503
1979	−0.207	0.115	232411	391720	304974
1980	−0.097	0.060	240549	419388	322795
1981	0.155	−0.051	248868	448633	341373
1982	−0.008	0.017	256780	478387	359766
1983	0.072	−0.033	263619	511328	379738
1984	−0.048	0.053	275283	543861	399254
1985	−0.273	0.150	289714	580258	421135
1986	−0.032	0.021	296989	617577	443259
1987	0.001	0.021	309918	659255	467511
1988	−0.057	0.084	330107	703388	493337
1989	0.055	0.003	345852	754820	523856
1990	0.076	−0.001	362567	810695	557323
1991	0.122	−0.022	377220	866834	590696
Aver.	0.001	0.040			

Source: OECD.

Note:

1. A'/A and B'/B are calculated using equations (6) and (7);
2. Y^1 is an estimate of Y calculated setting an initial value for the year 1909 and assuming Hicks neutral technical progress;
3. Y^2 is an estimate of Y calculated using biased technical progress as estimated in this table.

Table A9. Estimates of Growth Rates of Factor Augmenting Technical Progress and Output Assuming $\sigma = 0.6$.

	$\sigma = 0.6$				
	A'/A	B'/B	Y	Y^1	Y^2
1960			58275	58275	58275
1961	−0.051	0.145	65530	66719	64877
1962	0.003	0.055	71146	75639	71648
1963	−0.002	0.046	76940	85210	78721
1964	−0.016	0.123	86186	95521	86215
1965	0.013	0.014	90982	105962	93530
1966	−0.032	0.094	100860	118293	102118

(*Continued*)

Table A9. *(Cont'd)*

	$\sigma = 0.6$				
	A′/A	B′/B	Y	Y^1	Y^2
1967	−0.079	0.137	112498	132416	112075
1968	−0.068	0.157	127261	148784	123807
1969	−0.025	0.135	143298	166177	136358
1970	−0.055	0.122	158894	186078	151045
1971	−0.016	0.005	164740	204438	164413
1972	−0.024	0.079	177898	223714	178424
1973	0.012	0.041	191169	244379	193058
1974	0.068	−0.070	188292	258028	203085
1975	0.048	−0.020	191715	274743	214458
1976	0.003	0.014	200196	294920	226997
1977	0.023	0.018	209695	314970	239182
1978	−0.080	0.062	220307	335892	251493
1979	−0.084	0.066	232411	359511	265621
1980	−0.053	0.041	240549	381683	278993
1981	0.042	−0.002	248868	404966	292773
1982	−0.016	0.021	256780	428780	306092
1983	0.013	−0.010	263619	456320	320318
1984	−0.023	0.043	275283	481808	334027
1985	−0.104	0.083	289714	510154	349433
1986	−0.029	0.020	296989	538459	364783
1987	−0.004	0.023	309918	571626	381286
1988	−0.016	0.066	330107	604825	398960
1989	0.011	0.023	345852	642873	420162
1990	0.015	0.026	362567	681971	443640
1991	0.028	0.019	377220	719000	466850
Aver.	−0.016	0.051			

Source: OECD.

Note:

1. A′/A and B′/B are calculated using equations (6) and (7);
2. Y^1 is an estimate of Y calculated setting an initial value for the year 1909 and assuming Hicks neutral technical progress;
3. Y^2 is an estimate of Y calculated using biased technical progress as estimated in this table.

Table A10. Estimates of Growth Rates of Factor Augmenting Technical Progress and Output Assuming $\sigma = 0.5$.

	$\sigma = 0.5$				
	A′/A	B′/B	Y	Y^1	Y^2
1960			58275	58275	58275
1961	−0.042	0.139	65530	67001	64890
1962	−0.015	0.067	71146	76185	71647
1963	−0.016	0.054	76940	85966	78647
1964	−0.013	0.121	86186	96450	86066
1965	−0.009	0.027	90982	106985	93259
1966	−0.023	0.089	100860	119361	101648

TABLE A10. (*Cont'd*)

	$\sigma = 0.5$				
	A'/A	B'/B	Y	Y^1	Y^2
1967	−0.056	0.124	112498	133406	111430
1968	−0.048	0.144	127261	149464	122993
1969	−0.021	0.132	143298	166168	135371
1970	−0.049	0.118	158894	184835	149851
1971	−0.038	0.020	164740	201476	162956
1972	−0.029	0.082	177898	218573	176631
1973	−0.003	0.051	191169	236872	190846
1974	0.013	−0.038	188292	247262	200495
1975	0.012	−0.004	191715	261062	211405
1976	−0.005	0.018	200196	278895	223401
1977	0.010	0.023	209695	296343	234992
1978	−0.056	0.053	220307	314547	246634
1979	−0.060	0.056	232411	334939	259994
1980	−0.045	0.037	240549	353295	272586
1981	0.020	0.008	248868	372507	285501
1982	−0.018	0.022	256780	392307	297851
1983	0.002	−0.005	263619	416147	310868
1984	−0.017	0.041	275283	436998	323402
1985	−0.070	0.069	289714	460120	337487
1986	−0.029	0.020	296989	482708	351448
1987	−0.005	0.024	309918	510399	366215
1988	−0.008	0.062	330107	536831	382155
1989	0.002	0.027	345852	566765	401430
1990	0.002	0.032	362567	596090	422970
1991	0.009	0.027	377220	622455	444275
Aver.	−0.020	0.053			

Source: OECD.

Note:

1. A'/A and B'/B are calculated using equations (6) and (7);
2. Y^1 is an estimate of Y calculated setting an initial value for the year 1909 and assuming Hicks neutral technical progress;
3. Y^2 is an estimate of Y calculated using biased technical progress as estimated in this table.

Chapter 6

A NOTE ON MODELLING ENDOGENOUS GROWTH

1. Introduction

This note is to explore the conditions for the existence of a steady state in a neo-classical model with factor-augmenting technical progress. While many theoretical and empirical papers have been written elaborating on and testing the neo-classical model, the reason why Harrod-neutral technical progress alone is consistent with steady-state in these models has not been well understood or explained. This is particularly surprising as an earlier tradition in microeconomics has the relative growth in input prices determine bias endogenously.

Incorporating the Hicksian theory of bias in a neo-classical model with full employment poses many problems. Growth models traditionally assume that firms are atomistic; if firms are price takers, Fellner (1961) noted that some type of learning process will have to be postulated to justify such firms investing in biased technical innovations. Salter (1960) argued that firms should minimize total costs and should not be concerned with the cost or price of any one input; Kennedy (1964) responded to Salter's criticism by postulating a static innovation possibility frontier. Kennedy, however, took his theory to imply a rejection of the marginal productivity theory but Samuelson (1965) showed that it is consistent with the neo-classical theory of production [see also Drandakis and Phelps (1965)].

Nordhaus (1967) and Kamien and Schwartz (1969) developed microeconomic models in which the positions of the innovation possibility frontier are endogenously determined. But they did not consider biased technical progress. That step was taken in Sato and Ramachandran (1987). They showed that, in a steady state, the bias in technical progress will just counterbalance the differential growth in factor prices.

Economic historians would argue that, in industrialized countries, the rate of interest is oscillatory but trendless while the wage rate is steadily increasing. These trends in factor prices can be explained by a neo-classical model with Harrod-neutral technical progress if it is assumed that the economy is always in steady state; if the economy is not in steady state, the stylized facts are not

Acknowledgement. He wishes to acknowledge the helpful comments by Paul A. Samuelson, Rama Ramachandran, Chengping Lian, and the referee of this journal.

consistent with the model. The assumption that the economy is in steady state throughout the nineteenth and twentieth centuries is indeed a strong one. A model of endogenous bias can be made to generate the stylized facts even outside the steady state.

But the estimation of biased technical progress poses some well-known problems. The Diamond–McFadden theorem states that we cannot simultaneously estimate bias and elasticity. To break this impasse, Sato (1970) derived the Constant Elasticity of Derived Demand production function which has a convenient property that the elasticity of substitution is proportional to factor share. Estimation using U.S. non-farm data for 1909 to 1960 showed that the function fitted the data better than the Cobb-Douglas production function and that technical progress was labor saving. It also indicated that a rising trend in capital-labor ratio and constant interest rate and rising wage rate are not inconsistent. If $Y = F(AK,\ BL) = BL \cdot f(AK/BL) = BL \cdot f(\bar{k})$, where $\bar{k} = AK/BL$ then $\partial Y/\partial K = A(\mathrm{d}f/\mathrm{d}\bar{k}) = Af'(\bar{k})$ can be constant even if A is increasing provided f' decreases over time.

A model without technical progress can be formulated as

$$C = C(K,\ L\colon\ \dot{K},\ \dot{L}),$$

$$\frac{\partial C}{\partial K} > 0,\ \frac{\partial C}{\partial L} > 0,\ \frac{\partial C}{\partial \dot{K}} < 0 \text{ and } \frac{\partial C}{\partial \dot{L}} \leq 0,$$

where C, consumption, which is linear homogeneous in K, L, $\dot{K}$ and $\dot{L}$ and also satisfies other nice properties such as concavity. Here the economic assumption is that the growth of inputs, K and L, is achieved through the use of the homogeneous output so that consumption is determined not only by output but also by the rate of growth of K and L.

The standard neo-classical growth model assumes that the investment function is "linear," that is, $\partial C/\partial \dot{K} =$ constant, $\partial^2 C/\partial \dot{K}^2 = 0$ and $\partial C/\partial \dot{L} = 0$. Furthermore, saving, which is generated from output, is automatically invested and the labor force is exogenously given and growing rate at the rate n. Thus

$$C = F(K, L) - \dot{K},$$

$$\dot{K} = sF(K,\ L),\ 0 < s = \text{saving ratio} = \text{const.} < 1,$$

$$\dot{L} = nL. \quad n > 0.$$

Now introduce the factor-augmenting type of technical progress in production of output F as

$$F(A(t)K(t),\ B(t)L(t)),$$

where $A(t)$ and $B(t)$ are exogenously given and growing at the rate of α and β respectively, i.e.

$$A(t) = A(0)e^{\alpha t},$$
$$B(t) = B(0)e^{\beta t}.$$

Then the above model will become

$$C = F(A.K,\ B.L),$$
$$\dot{K} = sF(A.K,\ B.L),$$
$$\dot{L} = nL.$$

The capital accumulation function,

$$\frac{\dot{K}}{K} = sF\left(A,\ \frac{BL}{K}\right)$$

will be constant in steady state if and only if $F(A,\ BL/K)$ is constant. This is accomplished for a general class of F if and only if BL/K = constant and A = constant.[1] This means that in general the Harrod-neutral progress or the labor-augmenting type ($\alpha = 0,\ \beta \neq 0$) is the only case consistent with the steady state equilibrium. In steady state, the capital accumulation function

$$\frac{\dot{K}}{K} = sF\left(1,\ \frac{BL}{K}\right)$$

is homogeneous function of zero degree with respect to B, L, and K, and remains constant. This conclusion is valid even if we introduce "endogenous" labor augmenting technical progress as long as the rate of labor augmentation becomes constant in steady state, i.e.,

$$\lim_{t \to \infty} \frac{\dot{B}}{B} = \beta.$$

Next consider the case of "non-linear investment" function. To simplify the analysis, we consider the non-linear "separable" investment function.

$$C = F(K,\ L) - G(\dot{K},\ \dot{L},\ K,\ L).$$

Here F is the traditional production function. An example of a non-linear investment function in literature is [Leviathan and Samuelson (1969)]:

$$I^2 = (\dot{K} + \delta K)^2 + (\dot{L})^2 = G^2$$

and

$$C = F - G = F - I = F(1 - i)$$

where a fraction i of F allocated to create $\dot{K} + \delta K$ and $\dot{L}$.

This model can be extended to incorporate factor augmenting technical progress in the production function:

$$Y = F(AK,\ BL),$$
$$C = F(AK,\ BL) - G\cdot$$

[1] If F is Cobb–Douglas type or the limiting Cobb–Douglas type, where the function approaches to the Cobb–Douglas in the long run, then A need not be constant—a well-known result.

Just as the rate of growth of physical capital is modeled as being determined by a separable investment function, so it is traditional to assume (in models of endogenous technical progress) that the rates of growth of efficiency of inputs are determined by a separable "investment" function. In this paper, we generalize the function, $G(\cdot)$ above, to assume that it determines the growth rates of inputs in both physical and efficiency units through allocation of resources for "investment". Introducing technical progress in the factor augmenting form imposes very strong conditions on the properties of the investment and consumption functions. If the model is to have a steady state equilibrium in the sense that $\lim_{t\to\infty}(AK/BL)$ is a constant, G has to have functional properties which are compatible with that of F.

Note that $F(\cdot)$ is homogeneous of degree two in A, K, B and L. So **the necessary condition** for the steady state under the factor-augmenting technical progress is that $C = F(\cdot) - G(\cdot)$ is homogeneous of degree two; this would require that $G()$ is asymptotically homogeneous of degree two with respect to its variables, A, K, B, L, $\dot{K}$, $\dot{B}$, and $\dot{L}$. Many functions would satisfy this condition; some would have meaningful economic interpretation while others would not. The necessary condition can be written as:

$$\lambda^2 C = \lim_{t\to\infty} C(\lambda A.\lambda K, \lambda B.\lambda L; \lambda\dot{K}, \lambda\dot{L}, \lambda\dot{A}, \lambda\dot{B}, \lambda K, \lambda L, \lambda A, \lambda B) \quad (1)$$

An example of a function which satisfies this condition is,

$$C = F(AK,\ BL) - AK\left(\frac{\dot{A}}{A}\right)^2 - AK\left(\frac{\dot{K}}{K}\right)^2 - BL\left(\frac{\dot{B}}{B}\right)^2 \quad (2)$$

Since C is homogeneous of degree two in A, K, B, L, $\dot{K}$, $\dot{B}$ and $\dot{B}$, and A, K, B and L are growing exponentially at $t \to \infty$, there is a possibility that a steady state exists, with $\lim_{t\to\infty}(AK/BL) =$ constant.

The sufficient conditions: If *any* one of the following conditions is satisfied, then AK/BL is constant in steady state;

$$\lambda^2 C = C[\lambda A.\lambda K,\ \lambda B.\lambda L;\ \lambda(\dot{A}K),\ \lambda(\dot{B}L)] \quad (3)$$

or

$$\lambda^2 C = C[\lambda A.\lambda K,\ \lambda B.\lambda L:\ \lambda(\dot{A}K),\ \lambda(\dot{K}A),\ \lambda(\dot{B}L),\ \lambda(\dot{L}B),\ \lambda(\dot{A}\dot{K}),\ \lambda(\dot{B}\dot{L})] \quad (4)$$

or

$$\lambda^2 C = C\left[\lambda A.\lambda K, \lambda B.\lambda L; \lambda(AK)\phi_1\left(\frac{\dot{A}}{A}\right),\ \lambda(AK)\phi_2\left(\frac{\dot{K}}{K}\right),\ \lambda(BL)\phi_3\left(\frac{\dot{B}}{B}\right),\ \lambda(BL)\phi_4\left(\frac{\dot{L}}{L}\right)\right] \quad (5)$$

There may be many other forms of sufficient conditions but we will list below a few special functions that satisfy any one of the above sufficiency conditions.

(i) We now present three examples which satisfy the sufficient conditions (3), (4) and (5) respectively. The first case is a $G(\cdot)$ function that can be thought of as a generalization, to include technical progress, of the non-linear investment function stated earlier.

$$C = F(AK,\ BL) - G[(\dot{A}K),\ (\dot{B}L)] = (1 - i)F,\ G = iF \tag{6}$$

where $G^2 = a(\dot{A}K)^2 + b(\dot{B}L)^2$
Therefore,

$$C = F(AK,\ BL) - \sqrt{[a(\dot{A}K + A\dot{K})^2 + b(\dot{B}L + \dot{L}B)^2]}.$$

(ii) The following is an example of (4):

$$\begin{aligned} C = F(AK,\ BL) - G(.) = F(AK,\ BL) \\ - \sqrt{[\alpha(\dot{A}K)^2 + \beta(\dot{K}A)^2 + \gamma(\dot{B}L)^2 + \delta(\dot{L}B)^2 + \xi(\dot{A}\dot{K})^2 + \eta(\dot{B}\dot{L})^2]} \end{aligned} \tag{7}$$

(iii) A special case of (5) is the additive investment function of neoclassical growth models. The general form that will satisfy the homogeneity condition and the steady-state condition is

$$C = F(AK,\ BL) - \left[AK\phi_1\left(\frac{\dot{A}}{A}\right) + AK\phi_2\left(\frac{\dot{K}}{K}\right) + BL\phi_3\left(\frac{\dot{B}}{B}\right)\right]. \tag{8}$$

In particular, ϕ_i may take a power function form,

$$G = I = aAK\left(\frac{\dot{A}}{A}\right)^{\alpha} + bAK\left(\frac{\dot{K}}{K}\right)^{\beta} + cBL\left(\frac{\dot{B}}{B}\right)^{\gamma} \tag{9}$$

where $\alpha,\ \beta,\ \gamma > 1$.

An alternative implication of the sufficient condition may be more closely investigated by looking equation (8). The investment function $\dot{K}/K$, and the technical progress functions $\dot{A}/A$ and $\dot{B}/B$ must be of homogeneous degree zero with respect A, K, B and L. We have already shown that under the linear investment function $C = F - \dot{K}$, the capital accumulation function, $\dot{K}/K = sF(A,\ BL/K)$ is of homogeneous degree zero with respect to A, K, B and L, if and only if A is constant, i.e., the Harrod neutral technical change $\dot{B}/B = \beta > \beta$, and $\dot{L}/L = n$.

We can now formulate an optimal growth model with endogenous technical progress. Given $c = C/L = (F(AK,\ BL) - G(\dot{K},\ \dot{L},\ \dot{A},\ \dot{B},\ K,\ L,\ A,\ B))/L$ (Here we assume, without loss of generality, that the objective function can be written in a separable form)

$$\text{Max} \int_0^{\infty} e^{-\rho t} c\, dt. \tag{10}$$

Alternatively the above maximization problem may be written in the form of optimal control as

$$\underset{\{u_1,\, u_2,\, u_3\}}{\text{Max}} \int_0^{\infty} e^{-\rho t} c dt$$

s.t.

$$\dot{g}_1 = (u_1 + u_2 - \varepsilon) g_1$$
$$\dot{g}_2 = (u_3 + n - \varepsilon) g_2$$

where

$$c = CL^{-1}$$
$$= L^{-}.E.\left[F(g_1, g_2) - G\left(\frac{\dot{A}K}{E}, \frac{\dot{K}A}{E}, \frac{\dot{B}L}{E}, \frac{\dot{L}B}{E}, \frac{\dot{A}\dot{K}}{E}, \frac{\dot{B}\dot{L}}{E}, \frac{AK}{E}, \frac{BL}{E}\right)\right]$$
$$= e^{-nt} e^{\varepsilon t} [F(g_1, g_2) - G(u_1 g_1, u_2 g_1, u_3 g_2, n g_2, u_1 u_2 g_1, n u_3 g_2; g_1, g_2)]$$
$$g_1 = \frac{AK}{E},\ g_2 = \frac{BL}{E},\ \frac{\dot{A}}{A} = u_1,\ \frac{\dot{K}}{K} = u_2,\ \frac{\dot{B}}{B} = u_3,\ \frac{\dot{L}}{L} = n = \text{const.}$$

and

$$\frac{\dot{E}}{E} = \varepsilon = \frac{\dot{Y}}{Y} = \frac{(A^*\dot{K}^*)}{A^*K^*} = \frac{(B^*\dot{L}^*)}{B^*L^*} = \text{steady-state growth rate of output } Y$$

In order to solve the above problem, we follow the standard technique of optimal control. Setting

$$\underset{\{u_1,\, u_2,\, u_3\}}{\text{Max}} \int_0^{\infty} e^{-(\rho+n-\varepsilon)t} h(g_1, g_2, u_1, u_2, u_3) dt$$

subject to

$$\dot{g}_1 = (u_1 + u_2 - \varepsilon) g_1$$
$$\dot{g}_2 = (u_3 + n - \varepsilon) g_2$$

we obtain

$$H = h(g_1, g_2; u_1, u_2, u_3) + p_1 g_1 (u_1 + u_2 - \varepsilon) + p_2 g_2 (u_3 + n - \varepsilon)$$

The above can be solved to yield the optimal paths of investments in K, A and B [Sato, Ramachandran, and Lian (1993)].

In each period $t \geq 0$, the society chooses an optimal amount of investment I which is allocated among the three sectors that generate $\dot{A}/A$, $\dot{B}/B$ and $\dot{K}/K$. In a steady state, the investment ratio i takes a constant value of $i = i^*$.

This model can be worked out for the various special cases of G function that we have formulated above. As long as C is homogeneous of degree two, the system will have a steady state where the ratio AK/BL is a constant.

Finally we can show that the present model is applicable not only to a Robinson Crusoe economy, i.e., to the economy with an omnipotent capital planner, but also to a decentralized market economy in which consumers are responsible for intertemporal consumption planning and producers are seperately responsible for investment in physical and human capital and technical progress. Let us assume that a constant function of income is invested in $\dot{K}$, $\dot{A}$, and $\dot{B}$ respectively.

Thus, consider the model:

$$\begin{aligned}
&C = F(A.K,\ B.L) - \sum_{i=1}^{3} M_i \\
&M_i = s_i F(A.K,\ B.L),\ 0 < s_i < 1, \quad i = 1,\ 2,\ 3 \\
&M_1 = A.K\Phi_1\left(\frac{\dot{A}}{A}\right), \\
&M_2 = A.K\phi_2\left(\frac{\dot{K}}{K}\right) \quad \phi_i' > 0,\ \phi_i'' > 0 \\
&M_3 = B.L\Phi_3\left(\frac{\dot{B}}{B}\right) \\
&\dot{L} = nL
\end{aligned} \qquad (*)$$

The model reduces to the following:

$$\frac{\dot{Y}}{A} = \psi_1\left(\frac{s_1 F(A.K,\ B.L)}{A.K}\right), \quad \psi_1 = \Phi_1^{-1}$$

$$\frac{\dot{K}}{K} = \psi_2\left(\frac{s_2 F(A.K,\ B.L)}{A.K}\right), \quad \psi_2 = \Phi_2^{-1}$$

$$\frac{\dot{B}}{B} = \psi_3\left(\frac{s_3 F(A.K,\ B.L)}{B.K}\right), \quad \psi_3 = \Phi_3^{-1}$$

Defining $\bar{k} = A.K/B.L$, we have

$$\frac{\dot{\bar{k}}}{\bar{k}} = \psi_1(s_1 f(\bar{k})) + \psi_2(s_2 f(\bar{k})) - \psi_3(s_3 g(\bar{k})) - n,$$

where

$$f(\bar{k}) = F\left(1,\ \frac{1}{\bar{k}}\right) \text{ and } g(\bar{k}) = F(\bar{k},\ 1).$$

There exists a steady state equilibrium, i.e., $\dot{\bar{k}} = 0$ and $\bar{k} = \bar{k}^*$, and the economy is growing at the rate of

$$\left(\frac{\dot{A}}{A}\right)^* + \left(\frac{\dot{K}}{K}\right)^* = \left(\frac{\dot{B}}{B}\right)^* + n,$$

where * indicates the steady state value for each variable. The steady state euilibrium is stable, because

$$\frac{d\left(\frac{\dot{\bar{k}}}{\bar{k}}\right)}{d\bar{k}} = s_1\psi_1' f' + s_2\psi_2' f' - s_3\psi_3' g' < 0$$

where

$$\psi_i' > 0, \quad f' < 0 \text{ and } g' > 0.$$

We can show that the same stable property for a more general case where investment and technical progress functions satisfy the sufficient conditions stated above.

References

Drandakis, E. M. and Phelps, Edward S. "A model of induced innovation, growth and distribution", *The Economic Journal*, December 1965, **75**(300), pp. 823–40.

Fellner, William "Two propositions in the theory of induced innovations", *Economic Journal*, June 1961, **71**(283), pp. 305–8.

Kamien, Morton I. and Schwartz, Nancy. L. "Induced factor augmenting technical progress from a microeconomic viewpoint", *Econometrica*, September 1969, **37**(5), pp. 668–84.

Kennedy, Charles "Induced bias in innovation and the theory of distribution", Economic Journal, September 1964, **74**(295), pp. 541–47.

Leviathan, Nisan and Samuelson, Paul. A. "Notes on turnpikes: Stable and unstable", 1969, *Journal of Economic Theory*, December 1969, **1**(4), pp. 454–75.

Nordhaus, William D. *Invention, Growth, and Welfare: A Theoretical Treatment of Technological Change*, Cambridge, MA: The M.I.T Press, 1969.

Samuelson, Paul A. "A theory of induced innovation along Kennedy–Weizacker lines", *The Review of Economics and Statistics*, November 1965, **47**(4), pp. 343–56.

Salter, W. E. G. *Productivity and Technical Change, Cambridge*, U.K.: Cambridge University Press, 1960.

Sato, Ryuzo "The estimation of biased technical progress and production function", 1970, *International Economic Review*, June 1970, **39**(2), pp. 343–56.

Sato, Ryuzo and Ramachandran, Rama "Factor price variation and the Hicksian hypothesis: a microeconomic model", *Oxford Economic Papers*, June 1987, **39**(2), pp. 343–56.

Sato, Ryuzo, Ramachandran, Rama, and Lian, Chingpin "A model of optimal economic growth with endogenous bias", Working Paper of the Center for Japan-U.S. Business and Economic Studies, Stern School of Business, New York University, 1993.

Chapter 7

TECHNICAL CHANGE AND INTERNATIONAL COMPETITION

1. Introduction

Countries that are leaders in manufacturing technology in the post-World War II period, like US, Germany, France and UK, have generated scientific breakthroughs through endogenous research and development (R&D) activities. Latecomers like Japan, South Korea, and most of the developing countries have imported technology from the leaders. Technological leaders tend to invest a larger share of their R&D expenditures in basic research. The latecomers, on the other hand, invest relatively more in applied research and development. First by imitation, then by improvements on the processes and products imported, the latecomers may gain a competitive edge in export markets. They learn to produce the same (or similar) goods at a lower cost and export them to the world market. This aspect of international competition can best be explained in terms of dynamic comparative advantages of basic and applied innovations in the trading countries.

In my earlier paper (Sato, 1988), a differential game model was applied to study the basic nature of the international competition between the technological leaders and latecomers. The monopolistic firms in the two types of countries produce similar products and export them to the world market. These firms also engage in R&D activities: the firm in Country I (technologically advanced) engages in both basic and applied research activities, whereas the firm in Coun-

This paper was presented at the 16th Technical Symposium on "Differential Games: Applications to Economics and Finance", 31 March 2000, sponsored by The Center for Japan–US Business and Economic Studies, Stern School of Business, New York University.

try II imports basic technology from the firm in Country I and makes improvements to it in the form of process innovations. There are three relevant parameters which will affect the outcome of the competition for the market share of the two firms: (1) the index of diffusion of basic technology, (2) the index of relative efficiency of applied technology, and (3) the index of cost sharing of basic research.

It was shown in my earlier paper that the final outcome depends on the strategies adopted by these firms: the open-loop strategy or the closed-strategy. However, the analysis was incomplete. First, there may be several other outcomes in the open-loop case, and second, the closed-loop case must be completely re-evaluated.

The present paper attempts to analyze the two-country differential (asymmetric) game by re-evaluating both the open-loop and closed-loop strategies. It will be shown that the so-called "paradox" may emerge even in case of the open-loop strategy.

2. The Model

2.1. Production Function and Cost Function

There are two countries, Country I (USA) and Country II (Japan). Each country uses capital and labor to produce an identical good, y, with the same constant-returns-to-scale production function and Hicks neutral technical progress (process innovation),

$$y(t) = A(t)F[K(t), L(t)] \tag{1}$$

where K is the capital, L the labor and A is the level of Hicks neutral process innovation and the level of applied technology.

By solving the standard cost-minimization problem, the cost function is obtained from

$$\text{total cost function} = \frac{c(w, r)}{A(t)} y = Cy \tag{2}$$

where w and r are the wage rate and the return to capital, respectively. If w and r are constant, c is constant (set $c = 1$), and thus, we write $1/A = C$. Here C represents the level of the country's applied technology and the marginal cost of producing y.

2.2. Basic Research [Sato and Suzawa (1983)]

Only Country I engages in basic research, which is essential to produce applied technology. The basic research cost function is defined as

$$T = T(B, v) \tag{3}$$

where B is the level of basic technology and $v = \dot{B}$ is the time derivative of the basic technology. T is a convex function with the following characteristics:

$$\frac{\partial T}{\partial B} > 0 \tag{4.1}$$

$$\frac{\partial T}{\partial v} > 0 \tag{4.2}$$

Since Country II does not engage in basic research, it will import the basic technology from Country I, by sharing the cost of basic research. Let θ be the index of cost sharing which represents the percentage of total cost of basic research paid by Country II,

$$0 \leq \theta \leq 1 \tag{5}$$

2.3. Applied Research

Each country invests in applied research to reduce the marginal cost of producing output, C. The cost-reducing investment depends on the level of applied technology, the level of basic technology and the rate of change of the applied technology, i.e.

$$S = S(C, u, B), \; u = \dot{C} \tag{6}$$

where S is a convex function.

In addition the following properties are assumed on S:

$$\frac{\partial S}{\partial C} < 0 \tag{6.1}$$

$$\frac{\partial S}{\partial u} < 0 \tag{6.2}$$

$$\frac{\partial S}{\partial B} < 0 \tag{6.3}$$

Note that $C = 1/A$ and the condition (6.1) implies that as A increases (or C decreases), the cost of producing A will rise. Eq. (6.2) implies that the rate of increase in A will induce a higher cost to produce it. Finally, since the basic technology is a complement, an increase in B will help to reduce the cost of applied technology (6.3).

2.4. Quadratic R&D Cost Functions

In order to fully characterize the solutions of the asymmetric differential game, quadratic cost functions will be used. Thus,

$$T(B, v) = \frac{1}{2}\beta_{11}B^2 + \beta_{12}Bv + \frac{1}{2}\beta_{22}v^2 + \beta_0 \tag{7.1}$$

$$S(C, u, B) = \frac{1}{2}\alpha_{11}C^2 + \alpha_{12}Cu + \frac{1}{2}\alpha_{22}u^2 + \frac{1}{2}\alpha_{33}B^2 + \alpha_{13}CB + \alpha_{23}uB$$
$$- \alpha_1 C - \alpha_2 u - \alpha_3 B + \alpha_0 \tag{7.2}$$

where β_{ij}, α_{ij}, β_0 and α_i are positive constants.
In view of Eqs. (4.1), (4.2) and (6.1)–(6.3), S and T satisfy,

$$\begin{aligned}
\frac{\partial S}{\partial C} &= \alpha_{11}C + \alpha_{12}u + \alpha_{13}B - \alpha_1 < 0 \\
\frac{\partial S}{\partial u} &= \alpha_{12}C + \alpha_{22}u + \alpha_{23}B - \alpha_2 < 0 \\
\frac{\partial S}{\partial B} &= \alpha_{33}B + \alpha_{13}C + \alpha_{23}u - \alpha_3 < 0 \\
\frac{\partial T}{\partial B} &= \beta_{11}B + \beta_{12}v > 0 \\
\frac{\partial T}{\partial v} &= \beta_{12}B + \beta_{22}v > 0 \cdot
\end{aligned}$$

Assume that α_1, α_2 and α_3 are sufficiently large to ensure the inequalities, and $u = \dot{C}$ is usually negative when the firm invests to reduce the marginal cost, i.e. $\dot{A} > 0$ or $\dot{C} < 0$.

2.5. World Demand Function

The two firms are faced with the world demand function for the commodity y. Outputs of the two firms are perfect substitutes and the inverse demand function is linear,

$$P(Y) = P(y_1 + y_2) = a - b(y_1 + y_2),\ a > 0,\ b > 0 \tag{8}$$

where y_1 and y_2 are the output produced and sold in the world market by Countries I and II, respectively.

2.6. Efficiency of Applied Technology and Diffusion Index

The basic characteristic of the present model is that Country II gets the basic technology from Country I by the amount equal to

$$B_2 = \gamma B_1 \tag{9}$$

by paying θ percentage of the basic cost. If $\gamma = 1$, country receives the 100% level of the basic technology. There is no guarantee that this will be the case all the time. On the other hand, Country II can work harder to improve technology of the cost-reducing type S. It is assumed that Country II's applied technology is δ efficient (or inefficient), i.e. when $\delta \gtreqless 1$. Country II's applied technology is more

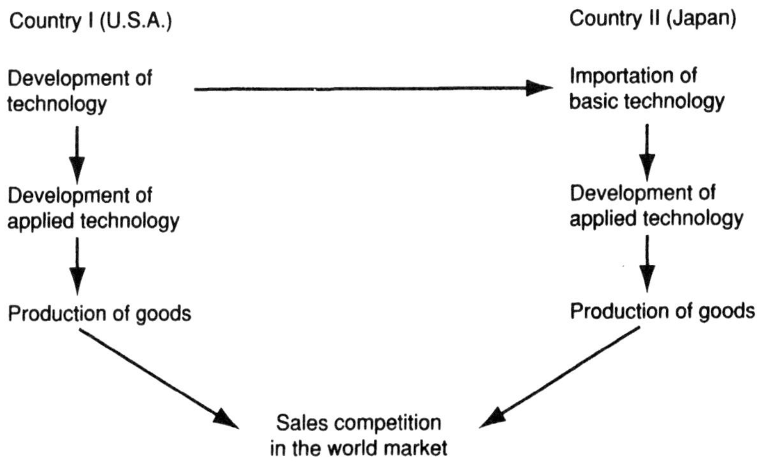

FIGURE 1. Structure of Technology Competition.

efficient, $\delta > 1$ (or less efficient $\delta < 1$) than that of Country I and, when $\delta = 1$, the two countries have the identical efficiency, thus

$$\delta \gtreqless 1 \tag{10}$$

In what follows $k = 1/\delta$ will be used for many instances. The model may be summarized by Fig. 1.

2.7. Output and Total Ggross Revenue

Given the world demand function and the marginal cost of producing y_i, it is straightforward to show that the output level y_i and total gross revenue R_i from the sales of y_i are given by equating the marginal revenue to the marginal cost as

$$y_i = \frac{a + C_j - 2C_i}{3b}, \quad i \neq j \tag{11}$$

and

$$R_i = \left(\frac{a + C_j - 2C_i}{3b}\right)^2, \quad i \neq j, \; i = 1, 2 \tag{12}$$

Obviously, the market share of each country is determined by the amount of marginal cost, C_i. In general, the output level y_i depends entirely on C_i. Thus,

$$y_1 \gtreqless y_2 \Leftrightarrow C_1 \lesseqgtr C_2 \tag{13}$$

2.8. Maximization Problem [Sato (1981)]

Each country seeks to maximize the long-run profit (total gross revenue minus R&D expenditures) by the appropriate choice of R&D investment.

Country I seeks to maximize,

$$\max_{u_1, u_2, v_1} \int_0^{\infty} e^{-\rho t} \times \left[\left(\frac{a + C_2 - 2C_1}{3b} \right)^2 - S(C_1, u_1, B_1) - (1 - \theta) T(B_1, v_1) \right] dt \quad (14.1)$$

subject to

$$\dot{C}_1 = u_1, \quad \dot{C}_2 = u_2(C_1, C_2, B_1), \quad \dot{B}_1 = v_1$$

Country II's problem is to

$$\max_{u_1, u_2, v_1} \int_0^{\infty} e^{-\rho t} \left[\left(\frac{a + C_1 - 2C_2}{3b} \right)^2 - kS(C_2, u_2, \gamma B_1) - \theta T(B_1, v_1) \right] dt \quad (14.2)$$

subject to

$$\dot{C}_1 = u_1(C_1, C_2, B_1), \quad \dot{C}_2 = u_2, \quad \dot{B}_1 = v_1(C_1, C_2, B_1)$$

where T and S are defined by Eqs. (7.1) and (7.2).

Notice that there are three parameters, θ, γ and $k = 1/\delta$, which represent the asymmetric nature of this model. It is this aspect of the model that makes the analysis more complicated but more interesting (and more relevant to reality) than the conventional two country model.

3. Technology game [Flaherty (1980)]

3.1. Open-loop Strategy

The result of open-loop competition in equilibrium may be summarized in the following propositions:

Proposition 1. *When Countries I and II are equally efficient in applied research* ($\delta = 1$, *or* $k = 1$) *and Country II receives full information of the basic technology* ($\gamma = 1$), *then in the steady-state the two countries have identical market shares,* $y_1 = y_2$. *The net profit of each country in the steady-state depends on the index of cost-sharing* θ. *When* $\theta = 1/2$, *the net profit of each country is identical.*

To prove the above proposition, we write k under the conditions $C_1 = C_2$ and $u_1 = u_2 = 0$ as

$$k = \frac{p - s}{p\gamma - s} \quad (15)$$

where

$$p = \alpha_3 X(n - z_2) - \alpha_{13} X(m - z_1) \quad (16.1)$$

$$s = X(\alpha_3 n - \alpha_{13} m) + z_1 n - z_2 m \quad (16.2)$$

$$X = \frac{\alpha_{13} + \rho\alpha_{23}}{\alpha_{33} + (1-\theta)(\beta_{11} + \rho\beta_{12})} \tag{16.3}$$

$$z_1 = \frac{4a}{3b} \tag{16.4}$$

$$z_2 = \frac{4}{3b} \tag{16.5}$$

$$m = \alpha_1 + \rho\alpha_2 \tag{16.6}$$

$$n = \alpha_{11} + \rho\alpha_{12}, \quad m > n \tag{16.7}$$

See Appendices A and B for details.

When $\gamma = 1$, $k = 1$, $\theta = 1/2$, the profit functions for Countries I and II (Eqs. (14.1) and (14.2)) are identical.

Proposition 2. Normal case

When, for any given θ,

$$p - s < 0 \text{ and } p > 0 \tag{17.1}$$

or

$$p - s > 0 \text{ and } p < 0 \tag{17.2}$$

then

$$\frac{\partial k}{\partial \gamma} > 0 \tag{18}$$

This is the case depicted in my original work as Fig. A.1 (1999, p. 202). The term $p - s$ can be expressed as

$$r = p - s = mz_2 - nz_1 - \alpha_3 X z_2 + \alpha_{13} X z_1 \tag{19}$$

For sufficiently small value of α_{13} (e.g. $\alpha_{13} = 0$), r in Eq. (19) will be negative, while p will be positive, provided that $n > z_2$ and $\alpha_3 > 0$.

Looking at n and z_2, we must have

$$n = \alpha_{11} + \rho\alpha_{12} > z_2 = \frac{4}{3b} \tag{20}$$

The above will be met when the absolute value of the slope of the demand function b is large, i.e. inelastic demand.

On the other hand, $r = p - s$ will be positive if m is sufficiently large relative to $z_1(\alpha_{13} \neq 0)$ and $n = z_2$, for $p < 0$. The normal case is depicted in Fig. 2.

Proposition 3. Paradox 1

When, for any given θ,

$$r = p - s < 0 \textit{ and } p < 0 \tag{21.1}$$

or

$$r = p - s > 0 \text{ and } p > 0 \tag{21.2}$$

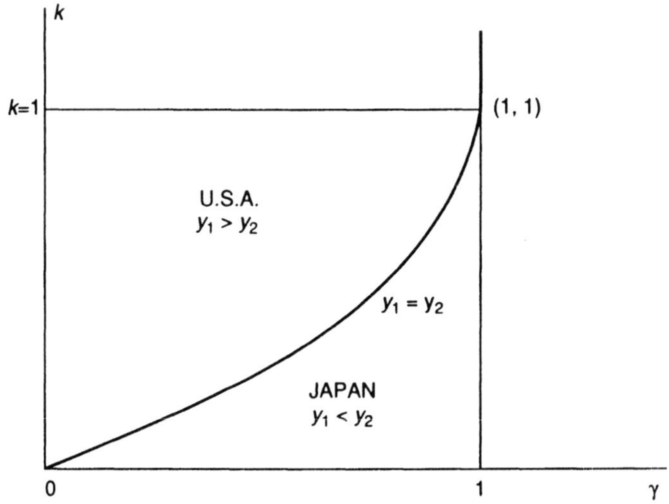

FIGURE 2. Normal Case.

then

$$\frac{\partial k}{\partial \gamma} < 0 \tag{22}$$

This is feasible under the conditions specified for S and T. In particular if

$$\alpha_3 > \alpha_{13}$$
$$\alpha_3 n \geq \alpha_{13} m$$

and

$$z_1 = z_2 = z \; (\Leftrightarrow a = 1)$$

then

$$p < 0 \text{ and } r = p - s < 0$$

provided that $m - n > 0$ is not very large.
On the other hand, when m is sufficiently large so that

$$r = p - s > 0$$

and

$$\alpha_3(n - z) > 0 \Leftrightarrow p > 0$$

then we have

$$\frac{\partial k}{\partial \gamma} < 0$$

This is a paradoxical case, because for $\gamma < 1$, we have $k > 1$. Fig. 3 depicts the paradox.

Remark 1. A simulation analysis shows that the paradoxical case, $\partial k/\partial \gamma < 0$, is the corner and limiting case where $y_1 = y_2; 0$. Hence, when $k > 1(\gamma = 1)$,

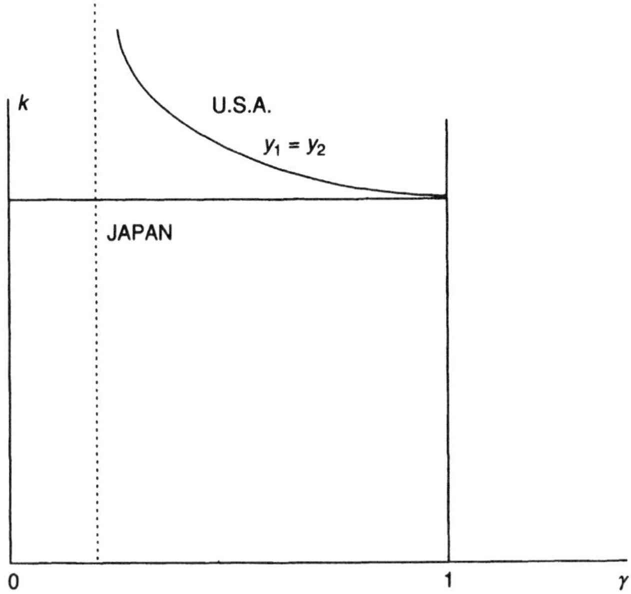

FIGURE 3. Paradoxical Case.

Country I monopolizes the world market, and when $k < 1(\gamma = 1)$, Country II monopolizes the market.

Remark 2. Under the assumptions that $\alpha_1 = 70$, $\alpha_2 = 80$, $\alpha_3 = 131$, $\alpha_{11} = 5.1$, $\alpha_{12} = 2$, $\alpha_{13} = 1$, $\alpha_{21} = 2$, $\alpha_{22} = 6$, $\alpha_{23} = 3$, $\alpha_{33} = 7$, $\alpha_0 = 3730$, $\beta_{11} = 9.85$, $\beta_{12} = 3$, $\beta_0 = -2550$, $\rho = 0.05$, $\theta = 0.5$, $a = 12$ and $b = 4$, the simulation shows that C_1 and $C_2 = 12.04104803$ and $y_1 = y_2 = -0.00342067$.

Proposition 4. Paradox 2

Consider the normal case (17.1). *Let $\theta = \theta_0$ be the value of θ which satisfies $r = 0$ or $p = s$, then as θ decreases beyond $\theta < \theta_0$, then the technology game has a paradox*

$$\frac{\partial k}{\partial \gamma} < 0$$

Using the example in Remark 2, we can show that $r = p - s = -0.2803$ and $p = 59.33303$ when $\theta = 0.6$, and $\gamma = k = 1$. When θ is reduced to $\theta = 0.5$, r changes to $r = 0.06527 > 0$ and $p = 54.38861 > 0$, which satisfies the condition (21.2). Fig. 4(a) and (b) illustrate this paradoxical case.

As long as $p > 0$, we can also show that as θ increases k will decrease, i.e.

$$\frac{\partial k}{\partial \theta} < 0$$

When $\theta \to \theta_0$ the normal case suddenly shifts to the paradoxical case from part (a) of Fig. 4 to part (b). This is illustrated by Fig. 5.

3.2. Closed-loop Strategies [Basar (1982)]

We will take up a special case of the closed-loop technology game. Assume that, as before output y_i is determined by 'the marginal revenue = the marginal cost'

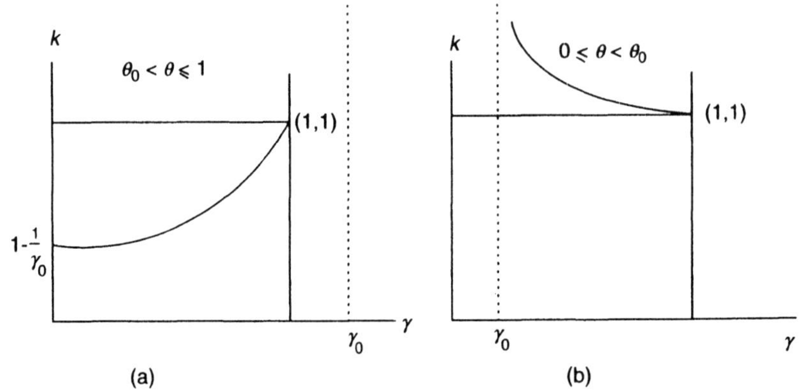

FIGURE 4. (a) Normal Case, (b) Paradoxical Case.

condition. Assume further that Country II is passive in developing the basic technology. That is to say, the basic technology is developed by Country I, but Country II pays no attention to the implicit cost of developing it. In short, Country II will not use the closed-loop strategy for the basic technology. Thus, we assume

$$\begin{aligned} \dot{C}_1 &= u_I = k_1 + m_1 C_1 + m_2 C_2 \\ \dot{C}_2 &= u_2 = k_2 + m_2^I C_I + m_I^I C_2 \end{aligned} \tag{23}$$

where m_{ij} are the coefficients of the closed-loop feedback mechanism. Unlike the symmetric game m_{ij}'s are not necessarily symmetric.

Proposition 5. Conjecture

When the two firms adopt the close-loop strategy, it is more likely that the competitive technology game will converge to the normal case.

Fig. 6 will illustrate this conjecture.

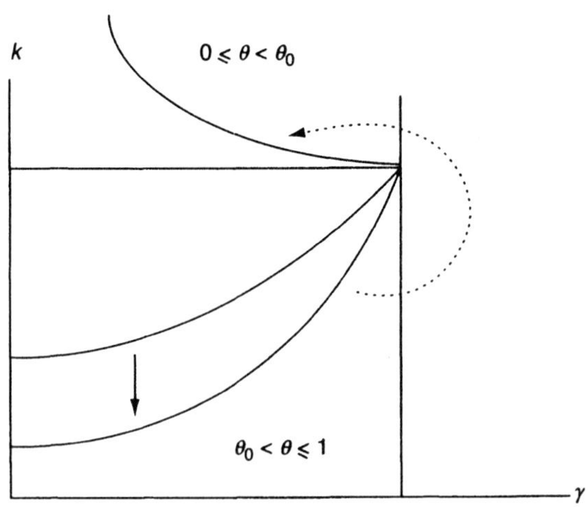

FIGURE 5. Shift of Normal Case to Paradoxical Case.

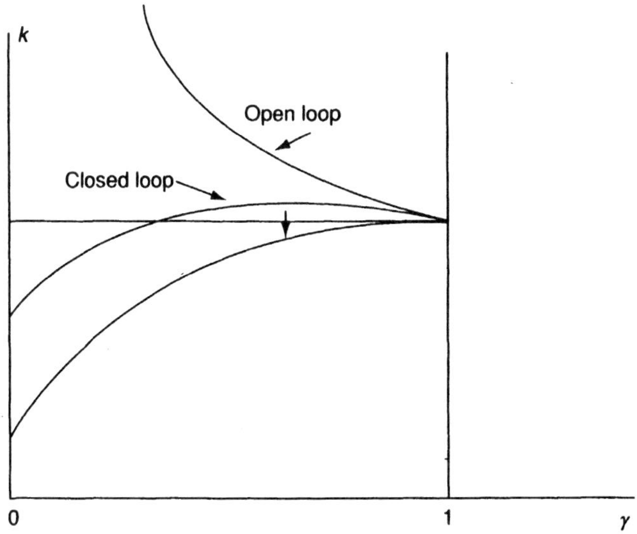

FIGURE 6. Closed-loop Convergence to Normal Case.

4. Concluding Remarks

We have developed a game-theoretic model of international technology competition. Country I (e.g. the US), a technologically mature country pursuing R&D in both basic and applied technology, competes in the world market with Country II (e.g. Japan) which specializes in research on applied technology. The model assumes that the technologically mature country has a dynamic comparative advantage to the extent that it can control the flow of information, or the diffusion index of the basic technology. The technologically less mature country may have a comparative advantage in applied technology if the flow of information is adequate.

The outcomes of the technology game depend on the strategies the two countries employ an open-loop or a closed-loop strategy. Under normal conditions, the open-loop strategy may give the two countries an equal market share in the world competition.

To put it differently, the technologically less advanced country can compete with the technologically more advanced country even though the flow of information of the basic technology is *not* adequate ($\gamma < 1$), provided that the former works harder to attain greater efficiency in the development of applied technology, $\delta > 1$.

A paradoxical case may occur when the technologically less advanced country *increases* the cost sharing of the basic technology ($\theta \to 1$). The country with no domestically produced basic research now can win the technology game even if its efficiency in applied technology is lower than that of the country producing basic research. To the technologically mature country, it appears that the other country is participating in "unfair competition".

The paradox may be used to analyze trade friction and fierce competition between the US and Japan in the post-World War II period. Japan sought to increase its market share, even by sacrificing its net profit.

The US, on the other hand, had a comparative advantage in basic research, but could not always utilize its advantage. The US criticized Japan for so-called "unfair" practice. The theoretical model developed in this paper may explain why Japan could win the game. It is not by "unfair" practice but by the nature of the technology game and by the use of basic research that the US developed.

The closed-loop strategy may eliminate the possibility of a paradox, by adjusting the control variables of each country. The precise mechanism, however, is not apparent in this complicated model of differential games with asymmetric parameters. The model must be dramatically simplified by combining both basic and applied research into one sector of R&D activities. This will be taken up as a future research project.

The simulation analysis also shows that the maximized profit of each country may not follow the maximized market share criterion. Thus, under certain conditions the country with a larger market share may not win a larger profit. This aspect of the simulation study will be reported in a separate research paper.

Acknowledgements

This paper is based on the joint research project "International Competition and Technology Game: Application of Asymmetric Differential Games" (by R. Sato, T. Nôno, F. Mimura and F. Fujiwara). We wish to express our appreciation to Paul Samuelson, Gilbert Suzawa, Norval Connell, Kunio Tanigaki, Michio Nikara and Hiroya Shimanuki for their helpful assistance.

Appendix A: Open-loop Strategy of a Technology Game Model "A Two-stage Maximization"

Country I's problem:

$$\max_{u_1, v_1, y_1} \int e^{-\rho t}[P(Y)y_1 - C_1 y_1 - S(C_1, B_1, u_1) - (1-\theta)T(B_1, v_1)]dt$$

subject to

$$\dot{C}_1 = u_1, \quad \dot{C}_2 = u_2, \quad \dot{B}_1 = v_1$$

Country II's problem:

$$\max_{u_2, y_2} \int e^{-\rho t}[P(Y)y_2 - C_2 y_2 - kS(C_2, \gamma B_1, u_2) - \theta T(B_1, v_1)]dt$$

subject to

$$\dot{C}_1 = u_1, \quad \dot{C}_2 = u_2, \quad \dot{B}_1 = v_1$$

Note that the integration is over the interval $[0, \infty]$.

In the above equations,

$$P(Y) = P(y_1 + y_2) = a - b(y_1 + y_2)$$

$$\begin{aligned} S(C_1, B_1, u_1) = & \frac{1}{2}\alpha_{11}(C_1)^2 + \alpha_{12}C_1u_1 + \frac{1}{2}\alpha_{22}(u_1)^2 + \frac{1}{2}\alpha_{33}(B_1)^2 + \alpha_{13}C_1B_1 \\ & + \alpha_{23}u_1B - \alpha_1C_1 - \alpha_2u_1 - \alpha_3B_1 + \alpha_0 \\ S(C_2, \gamma B_1, u_2) = & \frac{1}{2}\alpha_{11}(C_2)^2 + \alpha_{12}C_2u_2 + \frac{1}{2}\alpha_{22}(u_2)^2 + \frac{1}{2}\alpha_{33}(\gamma B_1)^2 + \alpha_{13}C_2(\gamma B_1) \\ & + \alpha_{23}u_2(\gamma B_1) - \alpha_1C_2 - \alpha_2u_2 - \alpha_3\gamma B_1 + \alpha_0 \end{aligned}$$

$$T(B_1, v_1) = \frac{1}{2}\beta_{11}(B_1) + \beta_{12}B_1v_1 + \frac{1}{2}\beta_{22}(v_1)^2 + \beta_0$$

First-stage maximization (open-loop strategy):
Hamiltonian

$$\begin{aligned} H^1 = & [a - b(y_1 + y_2)]y_1 - C_1y_1 - S(C_1, B_1, u_1) - (1-\theta)T(B_1, v_1) \\ & + \lambda_{11}u_1 + \lambda_{12}u_2 + \Pi_1v_1 \\ H^2 = & [a - b(y_1 + y_2)]y_2 - C_2y_2 - S(C_2, B_1, u_2) - \theta T(B_1, v_1) \\ & + \lambda_{22}u_2 + \lambda_{21}u_1 + \Pi_2v_1 \end{aligned}$$

$$\Rightarrow y_2 = \frac{a + C_1 - 2C_2}{3b}$$

$$\Rightarrow y_1 = \frac{a + C_2 - 2C_1}{3b}$$

Second-stage maximization:

$$\begin{aligned} \Rightarrow H^1 = & \frac{1}{9b}(a + C_2 - 2C_1)^2 - \left[\frac{1}{2}\alpha_{11}(C_1)^2 + \alpha_{12}C_1u_1 + \frac{1}{2}\alpha_{22}(u_1)^2 + \frac{1}{2}\alpha_{33}(B_1)^2\right. \\ & \left. + \alpha_{13}C_1B_1 + \alpha_{23}u_1B_1 - \alpha_1C_1 - \alpha_2u_1 - \alpha_3B_1 + \alpha_0\right] - (1-\theta)\left[\frac{1}{2}\beta_{11}(B_1)^2\right. \\ & \left. + \beta_{12}B_1v_1 + \frac{1}{2}\beta_{22}(v_1)^2 + \beta_0\right] + \lambda_{11}u_1 + \lambda_{12}u_2 + \Pi_1v_1 \\ \Rightarrow H^2 = & \frac{1}{9b}(a + C_1 - 2C_2)^2 - k\left[\frac{1}{2}\alpha_{11}(C_2)^2 + \alpha_{12}C_2u_2 + \frac{1}{2}\alpha_{22}(u_2)^2 + \frac{1}{2}\alpha_{33}(\gamma B_1)^2\right. \\ & \left. + \alpha_{13}C_2(\gamma B_1) + \alpha_{23}u_2(\gamma B_1) - \alpha_1C_2 - \alpha_2u_2 - \alpha_3(\gamma B_1) + \alpha_0\right] - \theta\left[\frac{1}{2}\beta_{11}(B_1)^2\right. \\ & \left. + \beta_{12}(B_1)v_1 + \frac{1}{2}\beta_{22}(v_1)^2 + \beta_0\right] + \lambda_{22}u_2 + \lambda_{21}u_1 + \Pi_2v_1 \end{aligned}$$

Impose steady-state

$$\Rightarrow B_1 = -\frac{\alpha_{13}C_1 - \alpha_3}{1-\theta}[\rho\beta_{12} + \beta_{11} + \alpha_{33}]$$

Let

$$\alpha_1 + \rho\alpha_2 = Y_1$$
$$\frac{\alpha_{13} + \rho\alpha_{23}}{\alpha_{33} + G} = X$$
$$\alpha_{11} + \rho\alpha_{12} = Y_2$$

$$G = (1-\theta)(\rho\beta_{12} + \beta_{11}) + \alpha_{33}$$
$$\Rightarrow C_1 = \frac{4a/3b - Y_1 + \alpha_3 X + 4C_2/3b}{8/3b - Y_2 + \alpha_{13}X}$$
$$\Rightarrow C_2 = \frac{4a/3b - kY_1 + k\gamma\alpha_3 X + C_1[4/3b - k\gamma\alpha_{13}X]}{8/3b - kY_2}$$
$$\Rightarrow C_1 = \frac{\begin{array}{c}16a/3b^2 - (4a/3b)kY_2 - (4/3b)(2+k)Y_1 \\ +(4/3b)(2+k\gamma)\alpha_3 X + kY_1Y_2 - kY_2\alpha_3 X\end{array}}{\begin{array}{c}16/3b^2 - (8/3b)(1+k)Y_2 + k(Y_2)^2 \\ +(4/3b)(2+k\gamma)\alpha_{13}X - kY_2\alpha_{13}X\end{array}}$$
$$\Rightarrow C_2 = \frac{\begin{array}{c}16a/3b^2 - (4a/3b)Y_2 - (4/3b)(2+k)Y_1 + (4/3b)(2+k\gamma)\alpha_3 X \\ +kY_1Y_2 - k\gamma Y_2\alpha_3 X + (4a/3b)(1-k\gamma)\alpha_{13}X\end{array}}{16/3b^2 - (8/3b)(1+k)Y_2 + k(Y_2)^2 + (4/3b)(2+k\gamma)\alpha_{13}X - kY_2\alpha_{13}X}$$

Recall $C_2 = C_1 = C_*$.

Note that the denominators of both $C_1 = C_2$ are equal.

$$\Leftrightarrow \left(\frac{4a}{3b} - Y_1 + \alpha_3 X\right)\left(\frac{4}{3b} - kY_2 + k\gamma\alpha_{13}X\right)$$
$$+ \left(\frac{4a}{3b} - kY_2 + k\gamma\alpha_3 X\right)\left(-\frac{4}{3b} + Y_2 - \alpha_{13}X\right) = 0$$

$\Leftrightarrow$ recall that $(\alpha_1 + \rho\alpha_2) = Y_1 = m,\quad \dfrac{\alpha_{13} + \rho\alpha_{23}}{\alpha_{33} + G} = X,\quad \alpha_{11} + \rho\alpha_{12} = Y_2 = n$

and let $z_1 = 4a/3b$ and let $z_2 = 4/3b$

$$\Leftrightarrow k[-z_1Y_2 - \alpha_3 Y_1 X + Y_1 z_2 + \alpha_{13}Y_1 X] + k\gamma[z_1\alpha_{13}X - \alpha_{13}Y_1 X - \alpha_3 X z_2 + \alpha_3 Y_2 X] + [-Y_1 z_2 + z_2\alpha_3 X + Y_2 z_1 - z_1\alpha_{13}X] = 0$$

Let

$$p = z_1\alpha_{13}X - \alpha_{13}Y_1X - z_2\alpha_3X + \alpha_3Y_2X = r + s,\ s = -Q$$
$$Q = -z_1Y_2 - \alpha_3Y_2X + Y_1z_2 + \alpha_{13}Y_1X$$
$$r = -[-Y_1z_2 + z_2\alpha_3X + Y_2z_1 - z_1\alpha_{13}X]$$
$$\Rightarrow pk\gamma + Qk = r$$
$$k[p\gamma + Q] = r$$
$$k = \frac{p-s}{p\gamma - s}$$

Appendix B: Closed-loop Strategy of a Technology Game Model

Country I's problem:

$$\max_{u_1, v_1, y_1} \int_0^{\infty} e^{-\rho t}[P(Y)y_1 - C_1 y_1 - S(C_1, B_1, \mu_1) - (1-\theta)T(B_1, v_1)]dt$$

subject to

$$\dot{C}_1 = u_1, \quad \dot{C}_2 = u_2(C_1, C_2, B_1), \quad \dot{B}_1 = v_1$$

Country II's problem:

$$\max_{u_2, y_2} \int_0^{\infty} e^{-\rho t}[P(Y)y_2 - C_2 y_2 - kS(C_2, \gamma B_1, u_2) - \theta T(B_1, v_1)]dt$$

subject to

$$\dot{C}_1 = u_1(C_1, C_2, \gamma B_1), \quad \dot{C}_2 = u_2, \quad \dot{B}_1 = v_1$$

Since the strategy is closed-loop we can set the following two equations: (i) $\dot{C}_1 = u_1$ and (ii) $\dot{C}_2 = u_2$, where

$$u_1 = K + m_1 C_1 + m_2 C_2, \quad u_2 = K + m_2^1 C_1 + m_1^1 C_2$$

For simplifying the Pontryagin necessary conditions for Countries I and II,

1. impose steady-state

$$\dot{\lambda}_{11} = \dot{\lambda}_{12} = \dot{\lambda}_{22} = \dot{\lambda}_{21} = FX \ldots\ldots = FX \ldots\ldots = u_1 = u_2 = v_1 = 0$$

2. set $\theta = 1/2$.

Let us now solve

$$C_2 = \frac{\begin{array}{c}[4a/3b - Y_1 - \alpha_3 X + (2m_2^1 a)/(3b(\rho - m_1))] \\ -C_1[-8/3b + Y_2 + \alpha_{13}X - (4m_2^1)/(3b(\rho - m_1))]\end{array}}{-[4/3b - (2m_2^1)/(3b(\rho - m_1))]}$$

$$C_2 = \frac{\begin{array}{c}4a/3b - kY_1 - k\gamma\alpha_3 X + (2m_2 a)/(3b(\rho - m_1)) \\ +C_1[4/3b + k\gamma\alpha_{13}X + (2m_2)/(3b(\rho - m_1))]\end{array}}{-8/3b + kY_2 - (4m_2)/(3b(\rho - m_1))}$$

Equating the above two expressions, we have

$$\Rightarrow \frac{\begin{array}{c}[4a/3b - Y_1 - \alpha_3 X + (2m_2^1 a)/(3b(\rho - m_1))] \\ -C_1[-8/3b + Y_2 + \alpha_{13}X - (4m_2^1)/(3b(\rho - m_1))]\end{array}}{-[4/3b - (2m_2^1)/(3b(\rho - m_1))]}$$

$$= \frac{\begin{array}{r} 4a/3b - kY_1 - k\gamma\alpha_3 X + (2m_2 a)/(3b(\rho - m_1)) \\ +C_1[4/3b + k\gamma\alpha_{13}X + (2m_2)/(3b(\rho - m_1))] \end{array}}{-8/3b + kY_2 - (4m_2)/(3b(\rho - m_1))}$$

We now solve for C_1.
Numerator:

$$\left[\frac{4}{3b} - \frac{2m_2^1}{3b(\rho - m_1)}\right]\left[\frac{4a}{3b} + \frac{2m_2 a}{3b(\rho - m_1)} - kY_1 - k\gamma\alpha_3 X\right]$$
$$- \left[-\frac{8}{3b} - \frac{4m_2}{3b(\rho - m_1)} + kY_2\right]\left[\frac{4a}{3b} + \frac{2m_2^1 a}{3b(\rho - m_1)} - Y_1 - \alpha_3 X\right]$$

Denominator:

$$\left[-\frac{8}{3b} - \frac{4m_2^1}{3b(\rho - m_1)} + Y_2 + \alpha_{13}X\right]\left[-\frac{8}{3b} - \frac{4m_2}{3b(\rho - m_1)} + kY_2\right]$$
$$- \left[\frac{4a}{3b} + \frac{2m_2}{3b(\rho - m_1)} - k\gamma\alpha_{13}X\right]\left[\frac{4}{3b} - \frac{2m_2^1}{3b(\rho - m_1)}\right]$$

$$C_1 = \frac{\begin{array}{r} [4a/3b - kY_1 - k\gamma\alpha_3 X + (2m_2 a)/(3b(\rho - m_1))] \\ -C_2[-8/3b + kY_2 - (4m_2)/(3b(\rho - m_1))] \end{array}}{4/3b + (2m_2)/(3b(\rho - m_1)) + k\gamma\alpha_{13}X_1}$$

$$C_1 = \frac{\begin{array}{r} 4a/3b - Y_1 - \alpha_3 X + (2m_2^1 a)/(3b(\rho - m_1)) \\ +C_2[4/3b - (2m_2^1)/(3b(\rho - m_1))] \end{array}}{-8/3b + Y_2 + \alpha_{13}X - (4m_2^1)/(3b(\rho - m_1))}$$

Equating the above two expressions, solve for C_2.
Numerator:

$$\left[\frac{4a}{3b} + \frac{2m_2^1 a}{3b(\rho - m_1)} - Y_1 - \alpha_3 X\right]\left[\frac{4}{3b} + \frac{2m_2}{3b(\rho - m_1)} + k\gamma\alpha_{13}X\right]$$
$$- \left[\frac{4a}{3b} + \frac{2m_2^1 a}{3b(\rho - m_1)} - kY_1 - k\alpha_3 X\right]\left[-\frac{8}{3b} - \frac{4m_2^1}{3b(\rho - m_1)} + Y_2 + \alpha_3 X\right]$$

Denominator:

$$\left[-\frac{8}{3b} - \frac{4m_2}{3b(\rho - m_1)} + kY_2\right]\left[-\frac{8}{3b} - \frac{4m_2^1}{3b(\rho - m_1)} + Y_2 + \alpha_{13}X\right]$$
$$- \left[\frac{4}{3b} - \frac{2m_2^1}{3b(\rho - m_1)}\right]\left[\frac{4}{3b} - \frac{2m_2^1}{3b(\rho - m_1)} + k\gamma\alpha_{13}X\right]$$

Note that the denominator of C_1 and C_2 is the same.
Equal market sharing $\Rightarrow C_1 = C_2$

$$\left[\frac{4}{3b}-\frac{2m_2^1}{3b(\rho-m_1)}\right]\left[\frac{4a}{3b}+\frac{2m_2a}{3b(\rho-m_1)}-kY_1-k\gamma\alpha_3X\right]$$
$$-\left[-\frac{8}{3b}-\frac{4m_2}{3b(\rho-m_1)}+kY_2\right]\left[\frac{4a}{3b}+\frac{2m_2^1a}{3b(\rho-m_1)}-Y_1-\alpha_3X\right]$$
$$=\left[\frac{4a}{3b}+\frac{2m_2^1a}{3b(\rho-m_1)}-Y_1-\alpha_3X\right]\left[\frac{4}{3b}+\frac{2m_2}{3b(\rho-m_1)}+k\gamma\alpha_{13}X\right]$$
$$-\left[\frac{4a}{3b}+\frac{2m_2^1a}{3b(\rho-m_1)}-kY_1-k\alpha_3X\right]\left[-\frac{8}{3b}-\frac{4m_2^1}{3b(\rho-m_1)}+Y_2+\alpha_3X\right]$$

References

Basar, T., Olsder, G.I., 1982. Dynamic Noncooperative Game Theory. Academic Press, New York (Revised Edition, 1995).

Flaherty, M.T., 1980. Industry structure and cost-reducing investment. Econometrica 48, 1187–1209.

Sato, R., 1981. Theory of Technical Change and Economic Invariance, Academic Press, New York (E. Elgar, Revised Edition, 1999).

Sato, R., 1988. In: Spence, A.M., Hazard, H.A. (Eds.), The Technology Game and Dynamic Comparative Advantage: An Application to US–Japan Competition in International, Competitiveness. Ballinger, Boston, MA, 1988, pp. 373–398 and Addendum (Revised Edition, and Reprinted as Chapter 16 in R. Sato, Production, Stability and Dynamic Symmetry, The Selected Essays of Ryuzo Sato, Vol. 2; E. Elgar, Economists of the Twentieth Century, 1999).

Sato, R., Suzawa, G., 1983. Research and Productivity: Endogenous Technical Progress. Greenwood Press, Wesiport, CT.

Part Two

ECONOMIC CONSERVATION LAWS

Chapter 8

OPTIMAL ECONOMIC GROWTH: TEST OF INCOME/WEALTH CONSERVATION LAWS IN OECD COUNTRIES

1. Introduction

For more than half a century, the theory of optimal economic growth has occupied a major position in modern economic analysis. Starting with the pioneering works of Ramsey (1928), Von Neumann (1945–1946), Samuelson and Solow (1956), Solow (1956), Cass (1965), and Sato and Davis (1971), to name a few, the theory of optimal economic growth has grown to include the more recent theory of "endogenous growth" (optimal or not optimal) of Paul Romer (1986), Robert Lucas (1988), and others. It is an old topic, but one that still presents new challenges for economists.

In this traditional approach, one major concern arises from determining the nature of appropriate choices in investment and how these decisions will influence the economy's path of long-run development. This method of analysis involves a typical application of "optimal" growth theory.

Another application of the optimal growth theory occurs when economists and statisticians attempt to justify and estimate the single index of "national

The author wishes to acknowledge the helpful comments by Paul A. Samuelson, Rama Ramachandran, William A. Barnett, and Changping Lian and the Statistical Assistance by Yukari Nakamura, Koji Hikosaka, and Youngduk Kim. Address correspondence to: Ryuzo Sato, C.V. Starr Professor of Economics and Director, Center for Japan-U.S. Business and Economic Studies, Stern School of Business, New York University, 44 West Fourth Street, New York, NY 10012-1126, USA; e-mail: rsato@stern.nyu.edu.

income" from a theoretical point of view. This problem has also attracted the attention of many economists, including Weitzman (1976), Sato (1981), Sato and Maeda (1990), and Kemp and Long (1992), among others. It has been shown that the best justification for national income (NNP or GNP) is the welfare measurement of the present discounted value of future consumption in an "optimally growing economy" [see Weitzman (1976)]. Although these two applications of the optimal growth theory appear to be unrelated, the problems of optimal capital accumulation and of calculating the correct and justifiable index of a nation's income can be seen as two sides of the same coin, that is, "optimality."

Attempts have been made to empirically test the validity of optimal growth models by comparing the simulated optimal time paths for the capital/labor ratio and the saving ratio with the actual time paths of these ratios [e.g., Sakakibara (1970) and Lenard (1972)]. Sakakibara and Lenard attempted to determine whether optimal growth models have any value as empirical tools for explaining economic development.

However, little empirical work has been done to test the validity of the second aspect of the growth theory, namely the testing of an appropriate measure of national income. The correct measure of NNP and GNP is considered "impractical" since wealthlike measures of the present discounted value of future consumption may not have an operational counterpart.

The purpose of this paper is to unify both aspects of the optimal growth theory with a general theory of "economic conservation laws." The theory of conservation law involves the identification and discovery of hidden invariant quantities in a dynamic system. In this dynamic economic system, as in a dynamic physical system, it is suspected that a certain variable remains unchanged during its process of evolution, as long as the system follows an optimal trajectory. In a growing economy, the variable that is invariant is called the conservation law [see Sato (1981)]. By uncovering the existence of conservation laws and by formulating the operational concepts associated with them, one can test the validity of both the optimal growth models as well as the measurement of national income. It has been shown that in an optimally controlled economy, the ratio of income to wealth remains invariant. This is the income/wealth conservation law.

1.1. Warning

The income/wealth conservation law states that the ratio of income to wealth should be constant over the entire period $0 \leq t \leq \infty$, if the economy is on an optimal path. This should not be confused with the standard concept of the income/capital ratio.

In the present model, like any other models of optimal growth, the income/capital ratio is not constant unless the economy is in the steady state. This conservation law guarantees that the income/wealth ratio is constant regardless of whether the optimally controlled economy is inside or outside the steady state.

1.2. Brief Summary

In the next section the basic model is presented. Conservation laws vary with the type of the objective function depicted, such as the maximization of the aggregate consumption or the maximization of per-capita consumption. It also depends upon the existence of exogenous factors, such as the exogenously growing labor force and/or technical change. In addition, the operational concept of "wealth-like quantity" is identified, although one should not always use the standard definition of "net national wealth" of Goldsmith and Kendrick (1976). The last section of the paper takes up an empirical analysis to determine how different economies have achieved long-term (optimal) growth. It indirectly verifies the correct and justifiable index of national income. We employ data from major OECD countries including the United States, Japan, Great Britain, Canada, and Germany.

Applications of various models illustrate that the U.S. economy has been operating rather efficiently along its optimal paths, despite the problems of business cycles and other maladjustments. The income/wealth ratio has remained almost constant for more than a century, with a distinct shift upward after World War II. On the other hand, the Japanese economy, after the oil shocks of the 1970's, has behaved rather differently. The income/wealth ratio has consistently declined, reaching the minimal value during the bubble period of the early 1990's. The economies of most of the other OECD countries, except those of Italy and Greece, have behaved similarly to that of the United States.

2. Model

Let consumption of the economy depend on output, which in turn depends on a vector of capital goods, a vector of investment and labor input, so that

$$C = C[Y(K;\ \dot{K};\ L;\ \dot{L})], \tag{1}$$

where $Y =$ output, which depends on $K = (K_1, \ldots, K_n) = n$ capital goods, $\dot{K} = (\dot{K}_1, \ldots, \dot{K}_n) = dK/dt =$ investment and L labor input, exogenously given by $\dot{L} = \lambda L$. [See Liviatian and Samuelson (1969).]

Let Y be a homogeneous function of the first degree with respect to its arguments $K, \dot{K}, L,$ and $\dot{L}$, together with $\partial Y/\partial K_i > 0$, $i = 1, \ldots, n$, $\partial Y/\partial L > 0$, $\partial Y/\partial \dot{K}_i, \leq 0$, $i = 1, \ldots, n$, $\partial Y/\partial \dot{L} \leq 0$. Also assume that $dC/dY > 0$ and Y is concave with respect to its argument. A special function of (1) is a separable case with $dC/dY = 1$.

$$C = Y(K, L) - G(\dot{K}, \dot{L};\ K, L) \tag{2}$$

or simply

$$C = Y(K, L) - G(\dot{K}, \dot{L}) \tag{3}$$

The simplest case of (3) is the standard model of saving = investment; that is,

$$C = Y(K, L) - \dot{K} \quad (i = 1). \tag{3a}$$

Another example of (3) may be a well-known form [(Caton and Shell (1971), Sato (1981), and Samuelson (1990)]

$$C = \prod_{i=1}^{n} K_i^{\alpha_i} L^{1-\Sigma\alpha_i} \left(\frac{1}{2} \sum_{i=1}^{n} \dot{K}_i^2 + \frac{1}{2} \dot{L}^2 \right)^{\frac{1}{2}} \tag{4}$$

with $1 > \alpha_i > 0$, $i = 1, \ldots, n$, and $\sum_{i=1}^{n} \alpha_i < 1$.

Without loss of generality, we can assume that $(dC)/(dY) = 1$ and (1) can be written as

$$C = F(K; \quad \dot{K}; \quad L; \quad \dot{L}). \tag{1'}$$

One of the distinct features of this model is that $\partial^2 F / \partial \dot{K}_i \partial \dot{K}_j \neq 0$. That is, the "investment function" is nonlinear, as shown by the example [equation (4)].

Consumption per capita is given by

$$\begin{aligned} c = \frac{C}{L} &= F\left(\frac{K_1}{L}, \ \ldots, \ \frac{K_n}{L}; \ \frac{\dot{K}_1}{L}, \ \ldots, \ \frac{\dot{K}_n}{L}; \ 1; \ \frac{\dot{L}}{L} \right) \\ &= F(k_1, \ \ldots, \ k_n; \ \dot{k}_1 + \lambda k_1, \ \ldots, \ \dot{k}_n + \lambda k_n; \ 1; \ \lambda), \end{aligned}$$

where

$$k_i = \frac{K_i}{L} \text{ and } \dot{k}_i = \left(\frac{\dot{K}_i}{K_i} - \lambda \right) k_i \quad i = 1, \ \ldots, \ n.$$

Thus we have

$$c = f(k_1, \ \ldots, \ k_n; \ \dot{k}_1 + \lambda k_1, \ \ldots, \ \dot{k}_n + \lambda k_n; \ \lambda) = f(k; \ \dot{k} + \lambda k; \ \lambda). \tag{5}$$

The society's objective is to maximize the discounted future value of consumption per capita, $c(t) = f(t)$, as

$$J = \int_0^{\infty} e^{-\rho t} f(k; \ \dot{k} + \lambda k; \ \lambda) dt \rightarrow \text{Max} \tag{6}$$

The necessary condition for the optimal solution is that the Euler–Lagrange equations vanish:

$$\begin{aligned} E_i &= \frac{\partial}{\partial k_i}(e^{-\rho t} f) - \frac{d}{dt}\left(\frac{\partial}{\partial \dot{k}_i}(e^{-\rho t} f) \right) \\ &= e^{-\rho t}\left[\frac{\partial f}{\partial k_i} + \rho \frac{\partial f}{\partial \dot{k}_i} - \frac{d}{dt}\left(\frac{\partial f}{\partial \dot{k}_i} \right) \right] = 0, \quad i = 1, \ \ldots, \ n. \end{aligned} \tag{7}$$

If we define the supply price of the ith capital per labor input as

$$P_i = -\frac{\partial f}{\partial \dot{k}_i}, \tag{8}$$

then the time derivative of the supply price of capital is

$$\dot{P}_i = -\frac{\partial f}{\partial k_i} + \rho P_i. \tag{8a}$$

It is very convenient to use the so-called Noether theorem [see Noether (1918), Nôno (1968), and Nôno and Mimura (1975, 1976, 1977, 1978)] to uncover both hidden and unhidden quantities working along the trajectory of the optimally controlled economy—known as the conservation law [Sato (1981, 1985), Sato, Nôno, and Mimura (1983), and Sato and Maeda (1990)]. Using the Noether theorem and its invariance principle, the general expression for the conservation law is

$$\Omega = \left(e^{-\rho t} f - e^{-\rho t} \sum_{i=1}^{n} \frac{\partial f}{\partial \dot{k}_i} \dot{k}_i \right) \tau + e^{-\rho t} \sum_{i=1}^{n} \frac{\partial f}{\partial \dot{k}_i} \xi^i - \Phi = \text{constant}, \tag{9}$$

where τ and ξ^i are the so-called infinitesimal transformations of t and k_i, respectively [see Appendix A, equation (A.7)].

In a special case of (9) in which $\tau = 1$, $\xi^i = 0, i = 1, \ldots, n, d\Phi/dt = -\rho e^{-\rho t} f$, we obtain the well-known conservation law of the income/wealth ratio as

$$f(k) - \sum_{i=1}^{n} \frac{\partial f}{\partial \dot{k}_i} \dot{k}_i = \rho \int_t^{\infty} e^{-\rho(s-t)} f(s) ds$$

or

$$c(t) + \sum_{i=1}^{n} P_i \dot{k}_i = \rho \int_t^{\infty} e^{-\rho(s-t)} f(s) ds \tag{10}$$

or

Consumption + Value of Investment per capita
$- \rho \times$ (Wealth measured in terms of per-capital future Consumption)

or

Income per capita $= \rho \times$ Wealth per capita

or

$$\rho = \frac{\text{Income per capita}}{\text{Wealth per capita}}. \tag{11}$$

We call the preceding the Per Capita Income/Wealth Conservation Law.

Note that equation (11) is *not* equal to (income per capita)/(capital per capita). Equation (11) is always constant for all, $0 \leq t \leq \infty$, while the income (per capita)/capital (per capita) ratio varies for $0 \leq t \leq \infty$ and is constant only at $t = \infty$, or in the steady state in the long run. [See Samuelson (1970) and Sato (1981).]

Because we are interested in time path of the aggregate economy, the above conservation law must be converted to the Aggregate Income/Aggregate Wealth

Law. This is not simply derived by multiplying both sides of (10) by L, because aggregate income must be defined by $C + \sum_{i=1}^{n} P_i \dot{K}_i$, which is not equal to $L(c + \sum_{i=1}^{n} P_i \dot{k}_i)$. In fact, $L(c + \sum_{i=1}^{n} P_i \dot{k})$ can be expanded as

$$L \cdot c + \sum_{i=1}^{n} P_i L\dot{k} = C + \sum_{i=1}^{n} P_i(\dot{K}_i - \lambda K_i), \tag{12}$$

where

$$P_i = -\frac{\partial C}{\partial \dot{K}} = -\frac{\partial c}{\partial \dot{k}}$$

and

$$L\dot{k}_i = L\frac{d}{dt}\left(\frac{K_i}{L}\right) = L\left[\frac{(\dot{K}_i L - \dot{L}K_i)}{L^2}\right] = \dot{K}_i - \lambda K_i.$$

However, because $\dot{K}_i - \lambda K_i$ is the adjusted net capital formation ($\dot{K}_i + \delta K_i$ being gross investment with $\delta =$ depreciation rate > 0, we call this an adjustment), we get the aggregate (adjusted) income/wealth conservation law as

$$C + \sum_{i=1}^{n} P_i(\dot{K}_i - \lambda K_i) = \rho e^{\lambda t} \int_t^{\infty} e^{-\rho(s-t)} f(s) ds \tag{13}$$

or

(Adjusted) National Income $= \rho \times$ National Wealth.

If we assume that "Labor" is also endogenous $K_{n+1} = L$, and the society's goal is to maximize the "aggregate" consumption, or if $\lambda = 0$, then (13) will reduce to the well-known result by Weitzman (1976) and Sato (1981, 1985).

From an empirical point of view, it is very important to formulate a model in such a way that the conservation laws can be tested against some observable data.

We have seen that when a factor such as labor is exogenously given to the system, maximization of consumption per labor input gives different conservation laws compared with the case of aggregate consumption maximization. Much of the literature in theoretical studies up to this point has basically ignored models of optimization with exogenous factors. As far as I know, there are only two exceptions: one where discount rate is changing over time and the other where there exists a kind of technical change (or taste change) in the system [see Sato (1985)]. These cases, in general, require a sophisticated data set to verify the underlying optimality conditions. We only take up a special case, the Harrod neutral type of technical change.

2.1. Harrod-Neutral Technical Change

If the aggregate consumption function is expressed as

$$C = F[K;\ \dot{K};\ BL;\ (B\dot{L})], \tag{14}$$

where BL = effective labor input with $B = B_0 e^{\beta t}$, $B_0 = 1$ and $\beta \geq 0$, or Harrod neutral technical change, then per-capita consumption is written as

$$c\frac{C}{L} = BF\left[\frac{K}{BL};\ \frac{\dot{K}}{BL};\ 1;\ \frac{(BL)}{BL}\right]$$
$$c = e^{\beta t} f[\bar{k};\ \dot{\bar{k}} + (\lambda + \beta)\bar{k};\ (\lambda + \beta)], \tag{15}$$

where

$$\bar{k} = (\bar{k}_1, \ldots, \bar{k}_n,), \dot{\bar{k}} = (\dot{\bar{k}}_1, \ldots, \dot{\bar{k}}_n) \text{ and } k_i = \frac{K_i}{BL}, \dot{\bar{k}}_i = \frac{d\bar{k}_i}{dt}, i = 1, \ldots, n.$$

The society's objective is to maximize

$$J = \int_0^\infty e^{-\rho t} c dt = \int_0^\infty e^{-(\rho - \beta)t} f[\bar{k};\ \bar{k} + (\lambda + \beta)\dot{\bar{k}};\ (\lambda + \beta)]dt, \tag{16}$$

where $\rho > \beta \geq 0$.

This is identical to (6) with ρ replaced by $(\rho - \beta)$, λ by $(\lambda + \beta)$, and k by $\bar{k}$, etc. Since the supply price of the ith capital

$$P_i = \frac{\partial F}{\partial \dot{K}_i} = \frac{\partial f}{\partial \dot{\bar{k}}}, \quad (i = 1, \ldots, n),$$

we obtain

$$c + \frac{1}{L}\sum_{i=1}^{n} P_i[\dot{K}_i - (\beta + \lambda)K_i] = (\rho - \beta)\int_t^\infty e^{-\rho(s-t)+\beta s} f(s)ds. \tag{17}$$

Again, by multiplying both sides of (17) by L, we get an aggregate expression for the conservation law:

$$C + \sum_{i=1}^{n} P_i[\dot{K}_i - (\beta + \lambda)K_i] = (\rho - \beta)e^{\lambda t}\int_t^\infty e^{-\rho(s-t)+\beta s} f(s)ds \tag{18}$$

$$\text{(Adjusted)National Income} = (\rho - \beta) \times \text{National Wealth.}$$

This is the modified conservation law; ρ is replaced by $(\rho - \beta)$ and the term expressing national wealth now contains $e^{\beta s} = B(s)$, as labor becomes more efficient by Harrod neutral technical progress. However, the conservation laws expressed by (13) and (18) are basically the same, which will be used for empirical applications.

3. How to Measure Wealthlike Present Value of the Stream of Future Consumption?

We next want to ask if there is any operational method and empirical counterpart to measure the wealthlike quantity $\int_t^\infty e^{-\rho(s-t)} C(s)ds$. We begin with the simplest case of $\lambda = 0$ and $\beta = 0$, or the endogenous labor force $K_n = L$. Then, the conservation law is

$$\text{Aggregate Income} = \rho \times \text{“Wealthlike Quantity.”}$$

Since the conservation law can be also written in (time) derivative form as [see Appendix A, equation (A.9)]

$$\frac{d}{dt}\left(C - \sum_{i=1}^{n} \dot{K}_i \frac{\partial F}{\partial \dot{K}_i}\right) = -\rho \sum_{i=1}^{n} \dot{K}_i \frac{\partial F}{\partial \dot{K}_i}, \tag{19}$$

where $C = F(K_1, \ldots, K_n; \dot{K}_1, \ldots, \dot{K}_n)$. Let $P_i = -\partial F/\partial \dot{K}_i$ and $Y = C + \sum_{i=1} P_i \dot{K}_i$, and then

$$\frac{dY}{dt} = \rho \sum_{i=1}^{n} P_i \dot{K}_i.$$

Then, by integrating both sides and utilizing (8a),

$$\dot{P}_i = -\frac{\partial F}{\partial K_i} + \rho P_i,$$

we have

$$\int_0^t dY(t) = \rho \int_0^t \sum_{i=1}^{n} P_i \dot{K}_i ds.$$

$$\begin{aligned} Y(t) - Y(0) &= \rho \left(\sum_{i=1}^{n} P_i K_i \right)_0^t - \int_0^t \sum_{i=1}^{n} \frac{d}{ds}[P_i(s)]K_i(s)ds \Big\} \\ &= \rho \left[V(t) - V(0) + \int_0^t \sum_{i=1}^{n} \frac{d}{ds}(F_{\dot{K}_i})K_i ds \right] \\ &= \rho \left[V(t) - V(0) + \int_0^t \sum_{i=1}^{n} \frac{\partial F}{\partial K_i} K_i ds - \rho \int_0^t \sum_{i=1}^{n} P_i K_i ds \right] \\ &= \rho \left[V(t) - V(0) + \int_0^t \prod (s)ds - \rho \int_0^t V(s)ds \right], \end{aligned} \tag{20}$$

where $V(t) = \sum_{i=1}^{n} P_i K_i(t) =$ value of capital or standard definition of wealth, $\prod(t) = \sum_{i=1}^{n} r_i K_i =$ total profit ($r_i =$ return space to K_i), and $(d/dt)(F_{\dot{K}_i}) = F_{K_i} + \rho F_{\dot{K}_i} V(t)$ is usually known as wealth in the national wealth literature. Our conservation law tells us that $Y \neq \rho V(t)$. Alternatively, we can present an operational expression from (19) as

$$\frac{dY}{dt} = \rho(Y - C) = \rho I. \tag{21}$$

By integrating, we immediately derive

$$Y(t) - Y(0) = \rho \int_0^t I(s)ds = \rho[W(t) - W(0)]. \tag{22}$$

The wealthlike quantity $\int_t^\infty e^{-\rho(s-t)}C(s)ds$ is exactly equal to $W(t)$. Hence the operational measure of wealthlike value of future consumption is nothing but the accumulated sum of the value of investments. In summary, we can use either (20), (21), or (22), because (20) requires ρ to be known a priori.

The next simplest case of maximization of utility of consumption per capita when $\lambda = 0$ requires several modifications. To apply aggregate data such as GNP or national wealth, we must use (13):

$$C + \sum_{i=1}^{n} P_i(\dot{K}_i - \lambda K_i) = \rho e^{\lambda t} \int_t^\infty e^{-\rho(s-t)} f(s)ds, \tag{13}$$

which can be expressed as

$$C(t) + P(\dot{K} - \lambda K) - L(t)y(0) = \rho L(t) \int_0^t \frac{I(s) - \lambda V(s)}{L(s)} ds. \tag{23}$$

The right-hand side represents the adjustments to GNP by $\lambda PK = \lambda V(t)$ and the initial condition $L(t)y(0)$, which are all identifiable from empirical data.

The above relationship can be applied to the case of Harrod neutral technical change by simply replacing λ with $\lambda + \beta$ and ρ with $\rho - \beta$ as equation (18) suggests. However, for the purpose of empirical estimation, both λ and β must be known before the above equation is tested for optimality.

In summary, we have presented two conservation laws. For each theoretical equation, there are at least three empirical counterparts. For the simplest of endogenous labor (or $\lambda = 0$), the three equations are (20), (21), and (22). In the empirical analysis presented in the next section, we used (20) and (21) to supplement and verify the value of ρ.

4. Test of Optimality and Conservation Laws

We begin with the application of the simplest optimal growth model (of per capita consumption maximization) for 12 OECD countries, including the United States, Canada, Japan, and Great Britain. However, as the first-order approximation we make *no* adjustments: that is, (a) no subtraction of $\lambda PK = \lambda V(t)$ on the left-hand side of (13) from GNP and (b) no adjustment in the exact calculation of "wealth" data [the right-hand side of (20)]. Instead, we simply use GNP data from the national income account and "net national wealth" data or value of capital goods in each country; that is,

$$Y = \text{GNP} \approx \rho \times \text{net national wealth} \approx \rho \times \sum_{i=1}^{n} P_i K_i = \rho V(t)$$

or

$$\frac{Y(t)}{V(t)} = \rho.$$

Obviously, when $\dot{P}_i$ is relatively close to zero, we have $V(t) \approx W(t)$. The results are remarkably consistent in general, and show that most economies can be

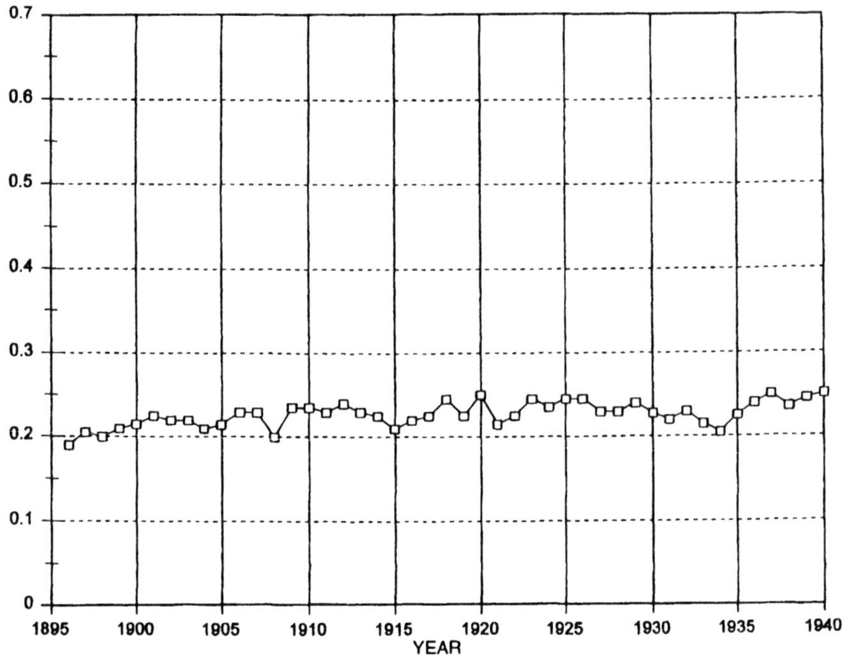

FIGURE 1. U.S. Income/wealth Ratio (Prewar).

viewed as operating along the optimal trajectories determined by the model. For instance, the U.S. economy for the most part showed a remarkable consistency in maintaining a relatively stable value of ρ around the value of 0.25 before the war (Figure 1) and around 0.3 in the post-World War II period (Figure 2). On the other hand, the Japanese economy behaved very differently from the U.S. economy in that ρ is consistently declining and approaches its lowest value during the bubble period of the early 1990's. This suggests that either the discount rate of the Japanese economy may be *variable* or that the Japanese economy may *not* be operating along the trajectories prescribed in the simple model. We take up each country separately.

5. Income/Wealth Ratio for the United States

The ratio of income to wealth for the United States has been historically stable, indicating that the income/wealth law may in fact be operating. No observable trend upward or downward occurs for the entire period 1896–1992 (see Figure 3). Note that, for a short period surrounding World War II (1941–1946), the ratios exhibit a pattern of sudden upward shift which is not quite consistent with the rest of the historical period under study and is not exhibited in the data point of the graphs. One important observation, however, is that this war period seems to indicate a structural change for the U.S. economy. Between the two periods before and after the war, there is a hysteretical break in the U.S. ratios, with the post-war ratios fluctuating approximately about a mean permanently higher than its prewar level.

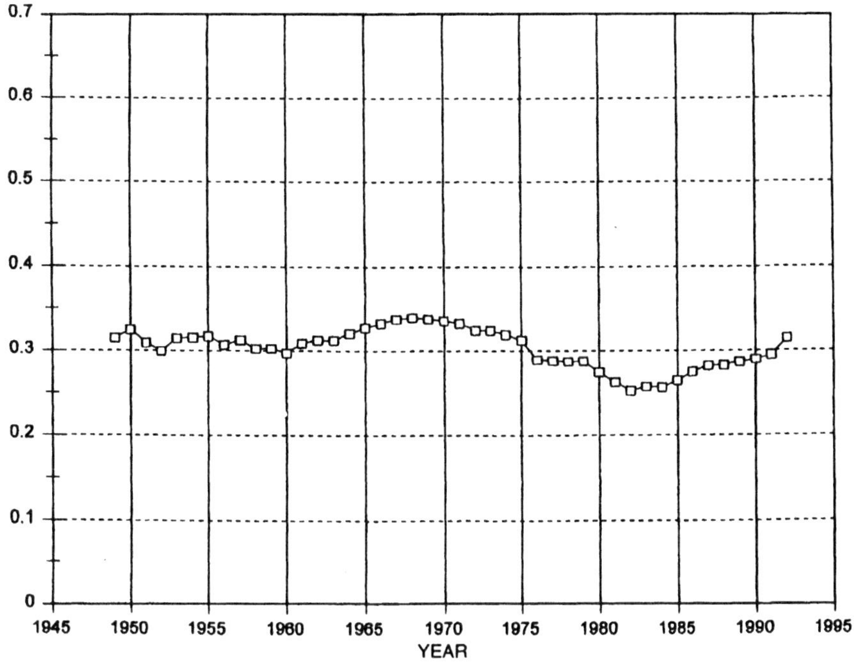

FIGURE 2. U.S. Income/wealth Ratio (Postwar).

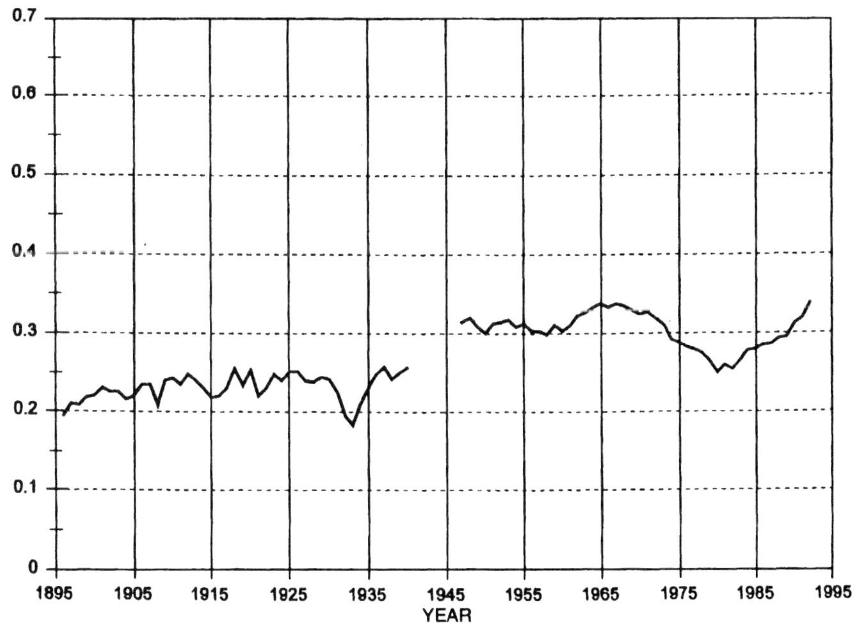

FIGURE 3. U.S. Ratio (GNP/national Private Wealth).

From the prewar period of 1896 and 1940, the U.S. ratio displayed stable fluctuations around its constant mean, although, during the Great Depression era of the early 1930's, income declined so that the ratio declined somewhat. The mean for this entire prewar period is $\rho = 0.23$, with variance 0.0003. In contrast, the mean for the postwar period 1947–1992 is higher at $\rho = 0.30$. Fluctuations about the postwar mean, however, are again stable with variance 0.0006.

Note that the postwar ratios display some business-cycle tendencies within their range. For the period immediately after the war and throughout the 1950's, the U.S. ratios are highly constant. This is followed in turn by a period in the 1960's that is largely of high ratios, due to a rise in income from the Vietnam War boom. The Ford-Carter era of the 1970's took the full shock of the oil crises, and, coupled with high asset price inflation, the ratio declined throughout the decade to a level below 0.3. It was not until Reagan's stimulative policy of the 1980's that the U.S. ratio turned upward again, and during the resultant Reagan boom of the decade, the ratio steadily climbed back. Meanwhile, the 1990's has seen a characteristic continuation in the rise of this ratio combined with a sharp decline in asset prices as the United States runs into a recession.

For the U.S. ratio, we have also studied cases of different levels of lag in wealth. One-year, two-year, and five-year lags were considered. Since none of these lags dramatically changes our results obtained without lags except that the ratios with lags tend to be lower, we will not include the results here. Those interested, however, may inquire about our results.

6. Statistical Tests for the United States

We have performed statistical tests to see if the conservations law holds for the United States and if there is autocorrelation in the disturbance. Our tests are based on earlier data derived from the President's Report (1992) covering the postwar period 1957–1990; see Appendix B.

To check for the conservation law, we first ran the following regression of GNP on net wealth, using 44 observations. The standard errors for the estimates are given inside the parentheses:

$$\begin{aligned} Y &= a + bW + \varepsilon, \\ a &= -13.5096(84.6638), \\ b &= 0.3331(0.0027). \end{aligned}$$

For the conservation law to hold, we require that the coefficient b equal the discount rate. This in turn requires that the constant term a be zero. Thus, we constructed a hypothesis,

$$\begin{aligned} H_0 &: a = 0 \\ H_A &: a \neq 0 \end{aligned}$$

Since the t-statistics in this case are equal to the ratio of the estimate for the constant term over its standard error, with the critical value for 5% significance level with 42 degrees of freedom being about 2.00, we can conclude that the null hypothesis is not rejected.

We next ran a regression on a specification without the constant term, and obtained the following result

$$Y = bW + \varepsilon$$
$$b = 0.3316(0.0019)$$

Thus, the estimated discount rate according to our conservation law is 0.3316 for the economy.

As for the autocorrelation test, our hypothesis was

$$H_0: r = 0$$
$$H_A: r \neq 0 \ (r\text{: correlation parameter})$$

For the sample size of 44 and 2 variables, the lower and the upper bounds of the Durbin–Watson distribution are 1.475 and 1.566, respectively. The obtained statistics equaled 1.61, which exceeded the upper bound. Therefore, we can conclude that the null hypothesis is not rejected and that there is no obvious autocorrelation in the disturbance term.

7. Income/Wealth Ratio for Japan

For the Japanese case, we used GDP data as income. For both GDP and net wealth, data series were taken directly from the original sources: Economic Planning Agency, Government of Japan, *Report on National Accounts from 1955 to 1989* for both GDP and net wealth data for the period 1955–1984; and the same agency's *Annual Report on National Account from 1993* for their continuing series for 1985–1991.

For the observed period of 1955 through 1991, the Japanese ratios are comparably lower than the U.S. ratios, with the mean equal to 0.21 and variance 0.0026 (Figure 4). The Japanese variance is higher than that of the United States because the Japanese ratios tend to exhibit a downward trend over time. The Japanese discount rate is considerably lower than that of the United States. This may be due to the fact that the Japanese are more long-sighted.

One similarity with the United States is a business-cycle-like movement of the Japanese ratio for the 1960's and 1970's, whereby high ratios are first observed for the 1960's and reflect a decade of high economic growth in Japan, which was then followed by a decade of decline in the ratios for the 1970's. Their rise and fall during these two decades seem to be contemporaneous with the U.S. case.

The real divergence in patterns comes in the 1980's, during which the Japanese ratio continues to decline while the U.S. ratio steadily rises. This persistent decline in the Japanese ratio coincides with the period of bubbles, during which asset prices skyrocketed two- or three-fold in Japan (see Figure 4).

Has the Japanese economy not grown along the optimal trajectories? Not necessarily. First, the standard model of a constant discount rate, $\rho(t) =$ constant, may not be relevant to the Japanese case. Second, because the capital price P_i may bediverging from its equilibrium price, i.e., $\dot{P}_k \neq 0$, much faster than in the United States, the true measure of the wealthlike quantity $W(t)$ and the value of capital $V(t)$ may be very different; most likely, $V(t)$ may be much greater than $W(t)$, especially during the bubble period. Further research will be undertaken in the near future.

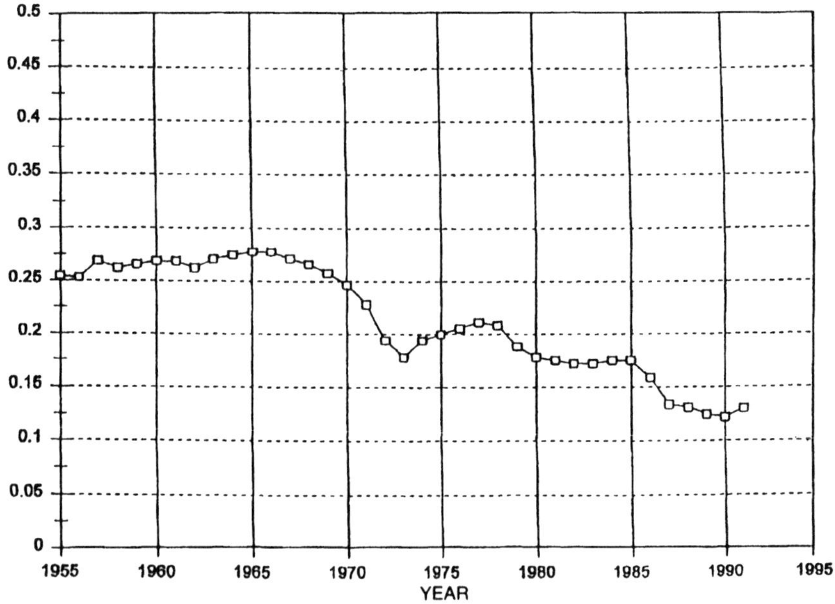

FIGURE 4. Ratio of GDP/net Wealth in Japan.

8. Income/Wealth Ratio for Other OECD Countries

Among the 24 OECD countries, we could collect data for 13 countries including the U.S. and Japan. The remaining 11 countries are: Australia, Canada, Finland, France, Germany, Greece, Iceland, Italy, Norway, Sweden, and the United Kingdom. The major data sources for these 11 countries is OECD, Department of Economics and Statistics, *Flows and Stocks of Fixed Capital: 1964–1989* (OECD: 1991).

Although we maintained Kendrick's concept of net wealth for OECD countries, because of a data availability problem, for some countries we were forced to use data that did not fully contain all components of net wealth as described in the section on U.S. wealth data. Also because of other data problems (i.e., observational sizes, data treatment procedures, etc., all slightly different across countries), our results are not for direct comparison between the levels of ratios across countries. In the following, therefore, we simply focus on one aspect: whether or not the constancy property of the ratio has been met. The 11 countries mentioned above can be roughly grouped into three categories: countries having constant ratios, those with declining ratios, and those with neither characteristic.

The first category consists of Australia, Canada, Norway, and Sweden (see Figures 5 through 8). The mean values of the ratios for their respective covered periods (see figures) are 0.35 for Australia, 0.31 for Canada, 0.27 for Norway, and 0.31 for Sweden. For these countries, the ratios seem to be highly constant over time. Thus, the income/wealth conservation law is present there.

On the other hand, Finland, Germany, Greece, and the United Kingdom make up the second group and have declining ratios (see Figures 9 through

12), a pattern similar to the Japanese case. Their respective means are 0.34 for Finland, 0.39 for Germany, 0.33 for Greece, and 0.38 for the United Kingdom.

Finally, Iceland and Italy follow neither of the above patterns (see Figure 13 and 14). For Iceland, the ratio is unstable, although there does not seem to be a trend, whereas for Italy the ratio goes up in the beginning of the series and comes down again. The means for those countries are 0.32 and 0.22, respectively.

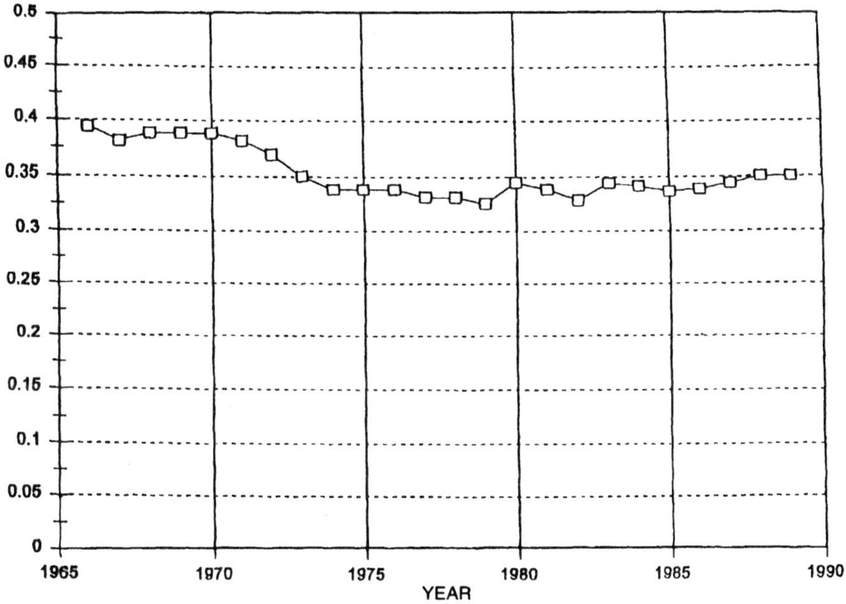

FIGURE 5. Australia.

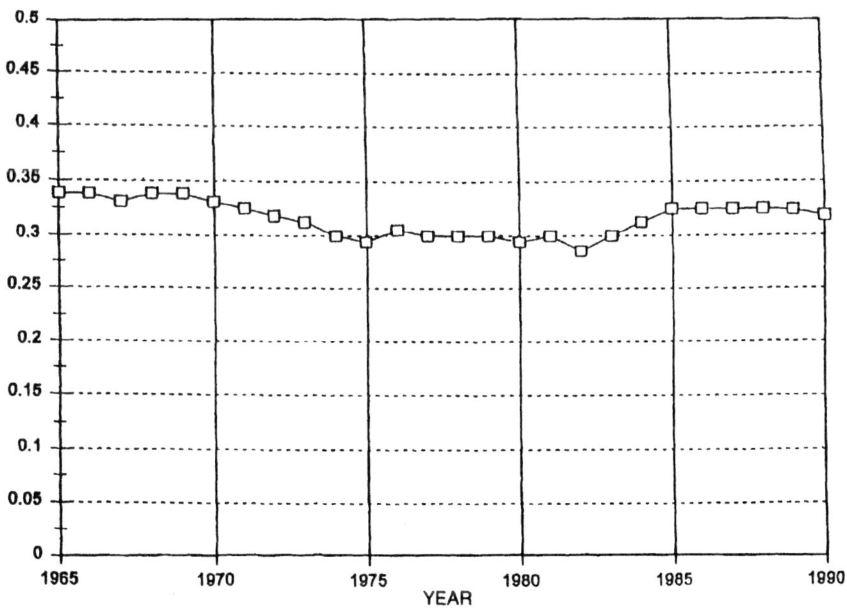

FIGURE 6. Canada.

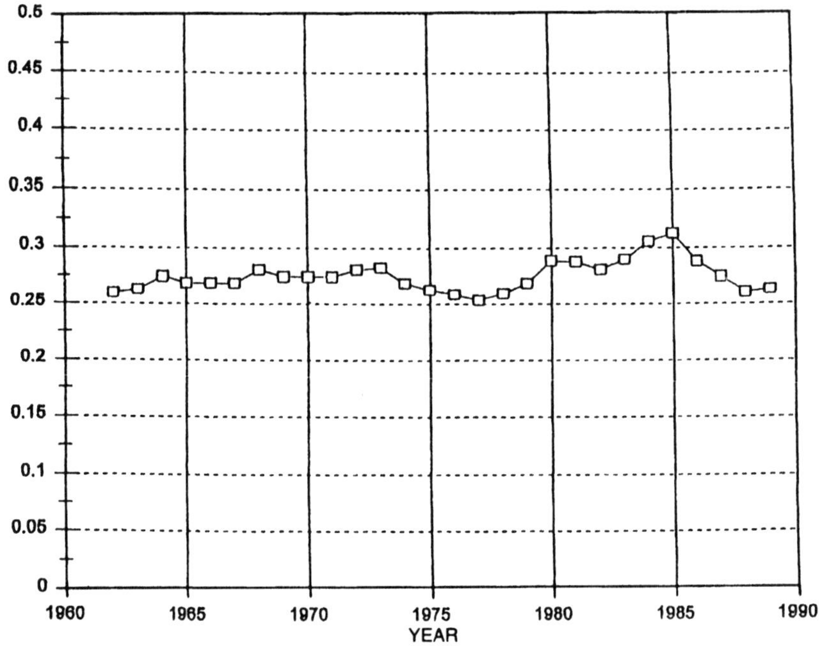

FIGURE 7. Norway.

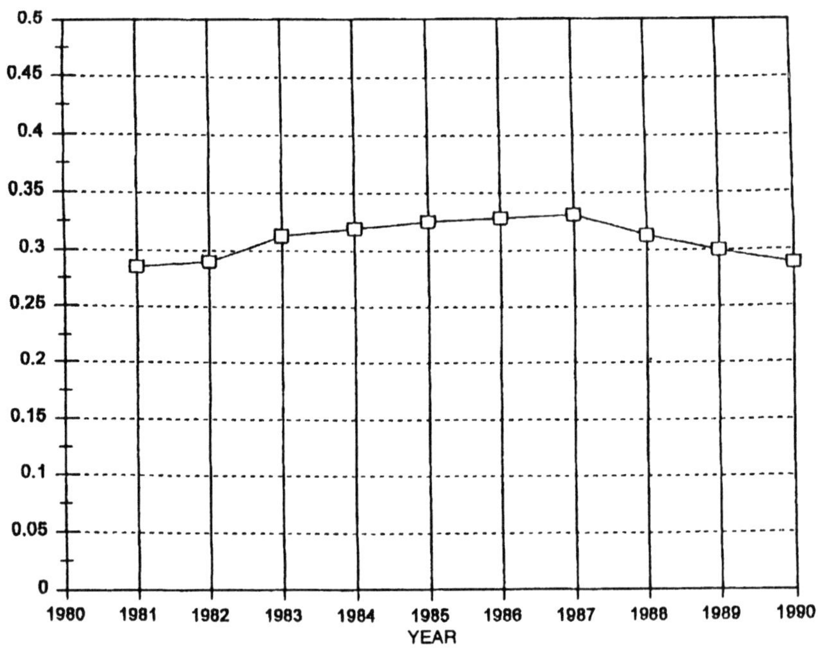

FIGURE 8. Sweden.

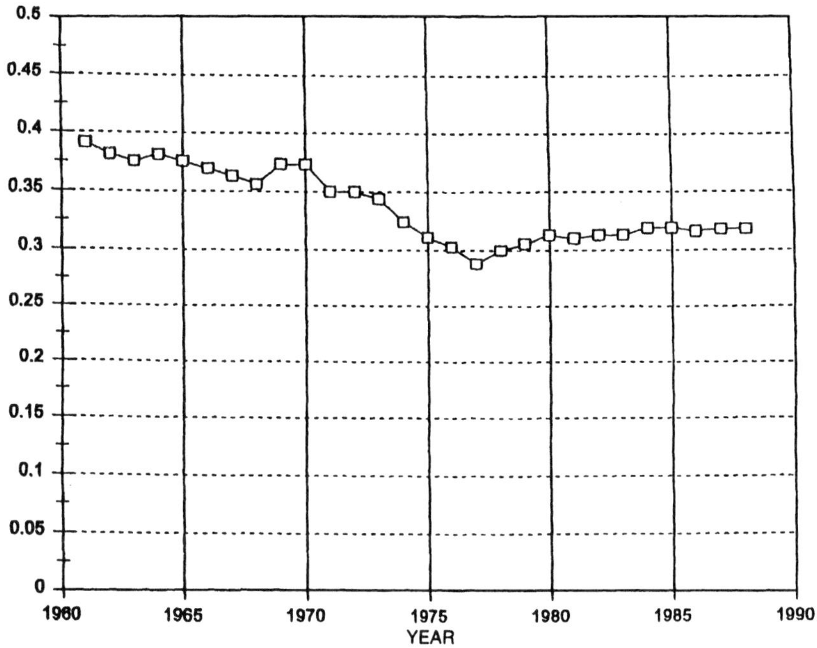

FIGURE 9. Finland.

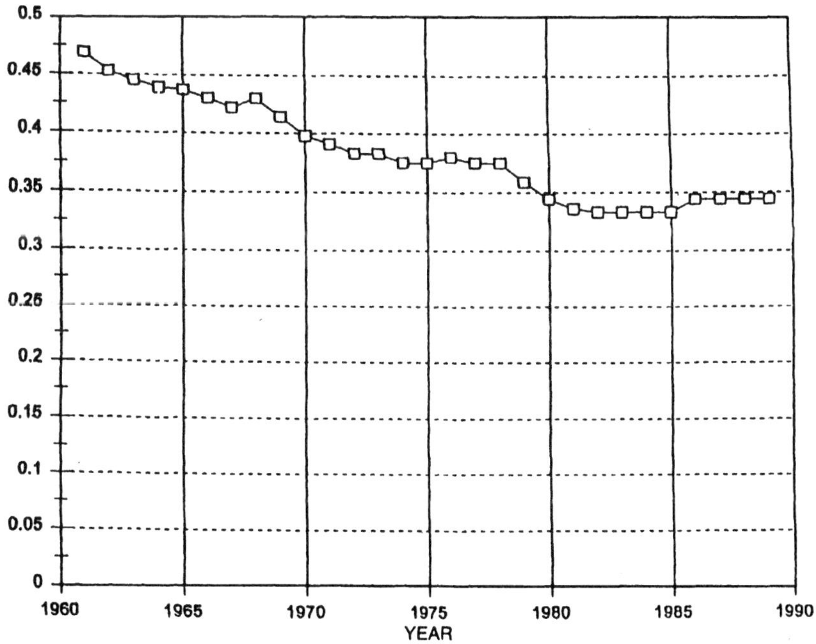

FIGURE 10. Germany.

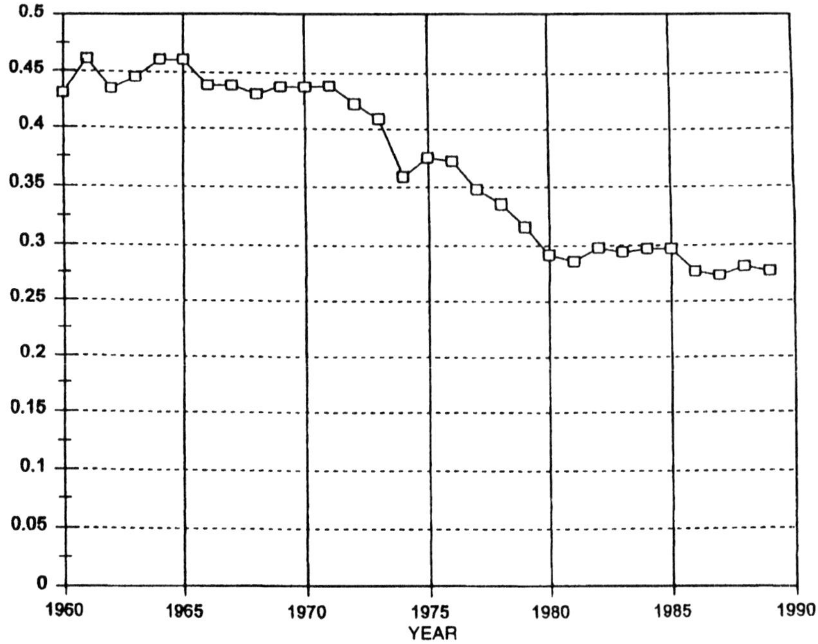

FIGURE 11. Greece.

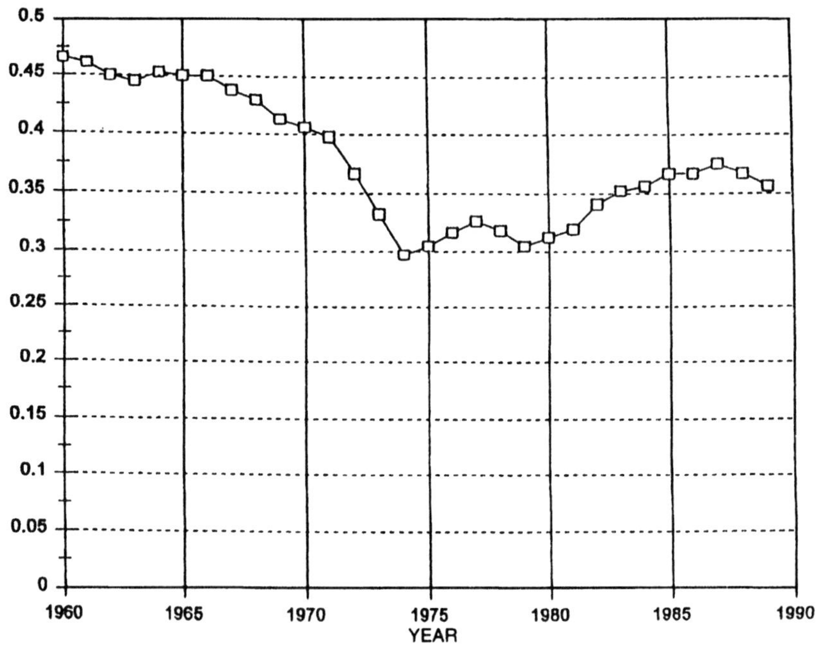

FIGURE 12. United Kingdom.

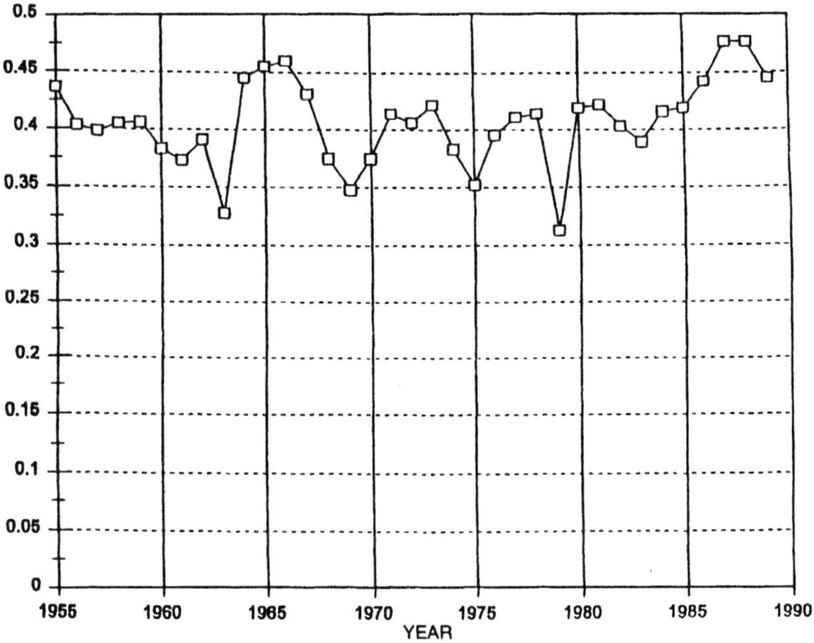

FIGURE 13. Iceland.

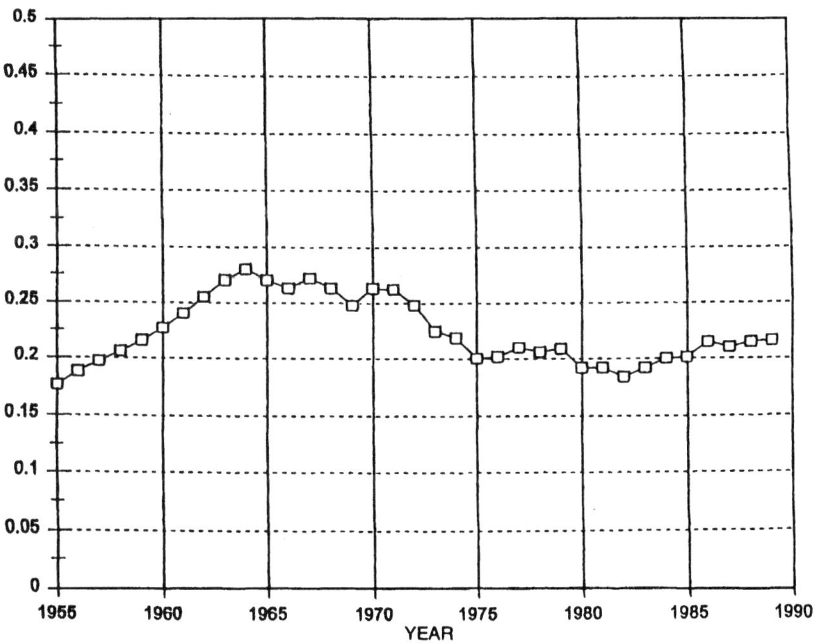

FIGURE 14. Italy.

References

Board of Governors of Federal Reserve System, *Balance Sheets for the U.S. Economy 1945–1992.*

Bureau of Economic Analysis (1992, 1993) *Survey of Current Business.* Washington, DC: U.S. Dept. of Commerce.

Cass, D. (1965) Optimum growth in an aggregative model of capital accumulation. *Review of Economic Studies* 32, 233–240.

Caton, C. & K. Shell (1971) An exercise in the theory of heterogeneous capital accumulation. *Review of Economic Studies* 38, 13–22.

Gelfland, I.M. & S.V. Fomin (1963) *Calculus of Variations.* Englewood Cliffs, NJ: Prentice-Hall, translated from the Russian by R.A. Silverman.

Goldsmith, R.W., *A Study of Saving in the United States,* vol. 3. New York: Greenwood Press.

Kemp, M.C. & N.V. Long (1992) On the evaluation of social income in a dynamic economy. In G.R. Feiwel (ed.), *Samuelson and Neoclassical Economics.* Boston: Kluwer-Nijhoff.

Kendrick, J. (1976) *The National Wealth of the United States.* New York: Conference Board.

Lenard, T.M. (1972) *Aggregate Policy Control and Optimal Growth: Theory and Applications to the U.S. Economy.* Ph.D. Dissertation, Brown University.

Lie, S. (1891) In G. Scheffers (ed.), Vorlesungen uber Differential gleichungen, mit bekanntel infinitesimalen Transformationen. Leipzig: Teubner; reprinted in 1967, New York: Chelsea.

Liviatian, N. & P.A. Samuleson (1969) Notes on turnpikes: Stable and unstable. *Journal of Economic Theory* 1, 454–475.

Logan, J.D. (1977) Invariant variational principles. In *Mathematics in Science and Engineering,* vol. 138. New York: Academic Press.

Lucas, R.E., Jr. (1988) On the mechanics of economic development. *Journal of Monetary Economics* 22, 3–42.

Moser, J. (1979) Hidden symmetries in dynamical systems. *American Scientist* 67, 689–695.

Noether, E. (1918) Invariante variations probleme, Nachr. Akad. Wiss. Gottingen, Math-Phys. KI-II, 235–257. Translated by M.A. Tavel (1971) Invariant variation problems. *Transport Theory and Statistical Physics* 1, 186–207.

Nôno, T. (1968) On the symmetry groups of simple materials: Application to the theory of Lie groups. *Journal of Mathematical Analysis and Applications* 24, 110–135.

Nôno, T. & F. Mimura (1975, 1976, 1977, 1978) Dynamic symmetries I/III/IV/V. *Bulletin of Fukuoka University of Education,* 25/26/27/28.

OECD, Department of Economics and Statistics (1991) Flows and Stocks of Fixed Capital: 1964–1989.

Ramsey, F. (1928) A mathematical theory of saving. *Economic Journal* 38, 54–559.

Romer, P.M. (1986) Increasing returns and long-run growth. *Journal of Political Economy* 94, 1002–1037.

Rund, H. (1966) *The Hamilton-Jacobi Theory in the Calculus of Variations.* Princeton, NJ: Van Nostrand-Reinhold.

Sagan, H. (1969) *Introduction to the Calculus of Variations.* New York: McGraw-Hill.

Sakakibara, E. (1970) Dynamic optimization and economic policy. *American Economic Review* 60.

Samuelson, P.A. (1970) Law of conservation of the capital-output ratio: Proceedings of the National Academy of Sciences. *Applied Mathematical Science* 67, 1477–1479.

Samuelson, P.A. (1990) Two conservation laws in the theoretical economics. In R. Sato & R. Ramachandran (eds.), *Conservation Laws and Symmetry.* Boston: Kluwer Academic.

Samuelson, P. & R. Solow (1956) A complete capital model involving heterogeneous capital goods. *Quarterly Journal of Economics* 537–562.

Sato, R. (1981) *Theory of Technical Change and Economic Invariance: Applications of Lie Groups.* New York: Academic Press (1981) (Reprint, Revised version, Cheltenham, UK and Northampton, MA: Elgar, 1999).

Sato, R. (1985) The invariance principle and income-wealth conservation laws. *Journal of Econometrics* 30, 365–389. Reprinted in *Production, Stability and Dynamic Symmetry, The Selected Essays of Ryuzo Sato,* vol. 2, Economists of Twentieth Century Series, Cheltenham, UK and Northampton, MA: Elgar, 1999, ch. 18.

Sato, R. & E.G. Davis (1971) Optimal savings policy when labor grows endogenously. *Econometrica* 39, 877–897. Reprinted in *Growth Theory and Technical Change, The Selected Essays of Ryuzo*

Sato, vol. 1, Economists of the Twentieth Century Series, Cheltenham, UK and Northampton, MA: Elgar, 1996, ch. 5.

Sato, R. & S. Maeda (1990) Conservation laws in continuous and discrete models. In R. Sato & R. Ramachandran (eds.), *Conservation Laws and Symmetry*. Boston: Kluwer Academic. Reprinted in *Production, Stability and Dynamic Symmetry, The Selected Essays of Ryuzo Sato*, vol. 2, Economists of Twentieth Century Series, Cheltenham, UK and Northampton, MA: Elgar, 1999, ch. 19.

Sato, R., T. Nôno & F. Mimura (1983) Hidden symmetries: Lie groups and economic conservation laws. Essay in honor of Martin Beckmann. Reprinted in *Production, Stability and Dynamic Symmetry, The Selected Essays of Ryuzo Sato*, vol. 2, Economists of Twentieth Century Series, Cheltenham, UK and Northampton, MA: Elgar, 1999, ch. 21.

U.S. Bureau of the Census (1975) *Historical Statistics of the U.S.*, Pt. 1. Washington, DC: U.S. Dept. of Commerce.

Von Neumann, J. (1945–1946) A model of general equilibrium. *Review of Economic Studies* 13, 1–9.

Weitzman, M.L. (1976) On the welfare significance of national product in a dynamic economy. *Quarterly Journal of Economics* 90, 156–162.

Appendix A

Consider the utility maximization problem, [See Sagan (1969), Gelfland and Fomin (1963), and Rund (1966)

$$\int_0^{\infty} U[t, k(t), \dot{k}(t)]dt \to \text{Max}, \tag{A.1}$$

where $k(t) = [k_1(t), \ldots, k_n(t)]$, $\dot{k}(t) = [\dot{k}_1(t), \ldots, \dot{k}_n(t)]$, and t in U represents exogenous factors such as labor force $L(t) = L_0e^{\lambda t}$, technical change and/or taste change.

Also consider r-parameter transformation [see Lie (1891) and Sato (1981)],

$$T: \bar{t} = \phi(t, k;\ \varepsilon), \varepsilon = (\varepsilon^1, \ldots, \varepsilon^r), \quad \bar{\kappa}_i = \psi^i(t, k;\ \varepsilon), \quad i = 1, \ldots, n, \tag{A.2}$$

where

$$\phi(t, k;\ 0) = t, \quad \psi^i(t, k;\ 0) = k_i, \quad i = 1, \ldots, n. \tag{A.3}$$

The infinitesimal transformations are given by

$$\tau_s(t, k) = \frac{\partial \phi}{\partial \varepsilon^s}(t, k;\ 0), \quad \xi_s^i(t, k) = \frac{\partial \psi^i}{\partial \varepsilon^s}(t, k;\ 0), \quad s = 1, \ldots, n,$$

and

$$X_s = \tau_s(t, k)\frac{\partial}{\partial t} + \xi_s^i(t, k)\frac{\partial}{\partial k_i} + \left(\frac{d\xi_s^i}{dt} - \dot{k}_i\frac{d\tau_s}{dt}\right)\frac{\partial}{\partial \dot{k}_i}, \tag{A.4}$$

where $\xi_s^i(t, k)(\partial/\partial k)$ is the Einstein notation for $\sum_{i=1}^n \xi_s^i(t, k)(\partial/\partial k_i)$, etc. Let (A.1) be given by

$$\int_0^{\infty} e^{-\rho t} U[k(t), \dot{k}(t)]dt \to \text{Max}. \tag{A.5}$$

Then, Noether's dynamic invariance condition with nullity Φ gives us [Noether (1918), Logan (1977), Sato (1981, p. 244, eq. 17)],

$$e^{-\rho t}\left[\rho U\tau + \frac{\partial U}{\partial k_i}\xi^i + \frac{\partial U}{\partial \dot{k}_i}\left(\frac{d\xi^i}{dt} - \dot{k}_i\frac{d\tau}{dt}\right) + U\frac{d\tau}{dt}\right] = \frac{d\Phi}{dt}$$

or

$$\left(U - \dot{k}_i\frac{\partial U}{\partial \dot{k}_i}\right)\frac{d\tau}{dt} = \rho U\tau - \frac{\partial U}{\partial k_i}\xi^i - \frac{\partial U}{\partial \dot{k}_i}\frac{d\xi^i}{dt} + e^{\rho t}\frac{d\Phi}{dt} \tag{A.6}$$

(since $s = 1$, $\tau_1 = \tau$, and $\xi_1^i = \xi^i$). The above will yield the conservation law

$$\Omega = \left(e^{-\rho t}U - e^{-\rho t}\dot{k}_i\frac{\partial U}{\partial \dot{k}_i}\right)\tau + e^{-\rho t}\frac{\partial U}{\partial \dot{k}_i}\xi^i - \Phi = \text{const.} \tag{A.7}$$

or

$$\frac{d\Omega}{dt} = -\rho e^{-\rho t}\left(U - \dot{k}_i\frac{\partial U}{\partial \dot{k}_i}\right)\tau + e^{-\rho t}\frac{d}{dt}\left[\left(U - \dot{k}_i\frac{\partial U}{\partial \dot{k}_i}\right)\tau\right]$$

$$+e^{-\rho t}\left(U - \dot{k}_i\frac{\partial U}{\partial \dot{k}_i}\right)\frac{d\tau}{dt} - \rho e^{-\rho t}\frac{\partial U}{\partial \dot{k}_i}\xi^i$$

$$+e^{-\rho t}\frac{d}{dt}\left(\frac{\partial U}{\partial \dot{k}_i}\xi^i\right) - \frac{d\Phi}{dt} = 0 \tag{A.8}$$

By eliminating $d\Phi/dt$ between (A.6) and (A.8) and setting $\xi = 0$, $\tau = 1$, we obtain our first "income/wealth" conservation law as

$$\frac{d}{dt}\left[U - \dot{k}_i\frac{\partial U}{\partial \dot{k}_i}\right] = -\rho\dot{k}_i\frac{\partial U}{\partial \dot{k}_i} \tag{A.9}$$

or

$$\frac{d}{dt}(\text{income at } t) = \rho \times (\text{Utility Value of Investment at } t),$$

where $-\partial U/\partial \dot{k}_i = P_i =$ price of capital. Also, from (A.6) and (A.8), we have

$$\frac{d}{dt}\left[e^{-\rho t}\left(U - \dot{k}_i\frac{\partial U}{\partial \dot{k}_i}\right)\right] = -\rho e^{-\rho t}U. \tag{A.10}$$

Integrating the above,

$$\Phi|_t^\infty = e^{-\rho s}\left\{\left[U(s) - \dot{k}_i(s)\frac{\partial U(s)}{\partial \dot{k}_i(s)}\right]\right\}_t^\infty = -\rho\int_t^\infty e^{-\rho s}U(s)ds$$

and using the transversality condition $\Phi(\infty) = 0$, we have

$$e^{-\rho s}\left[U(t) - \dot{k}_i(t)\frac{\partial U(t)}{\partial \dot{k}_i(t)}\right] = \rho \int_t^{\infty} e^{-\rho s} U(s)ds,$$

which will reduce to Weitzman's original result [Weitzman (1976)],

$$\text{Income} = U(t) - \dot{k}_i(t)\frac{\partial U(t)}{\partial \dot{k}_i(t)} = \rho \int_t^{\infty} e^{-\rho(s-t)} U(s)ds; \qquad \textbf{(A.11)}$$

that is,

$$\text{Income} = \rho \times \text{``Wealthlike Quantity''} \qquad \textbf{(A.12)}$$

[see Sato (1985), Sato and Maeda (1990)].

Appendix B

B.1. Notes on U.S. Wealth (Net worth) Data

For the purpose of our study, we followed the John Kendrick's (1976), definition of wealth in which wealth is taken as the stock of productive capacity resulting from past investments. Furthermore, Kendrick confines wealth to the conventional tangible assets not including human capital.

Domestic financial assets do not come into play for our purposes because consolidation of all domestic assets results in cancellation of domestic financial claims and assets held by all sectors of the economy—households, corporations, and the government—because every domestic claim is a liability at the same time within the national boundaries. The domestic net worth therefore reduces to the total amount of tangible assets held in the domestic economy. By adding to this the net surplus in foreign assets held by the domestic sectors, one will obtain national wealth. Following this set of definitions thus leads to wealth and net worth being synonymous at the national level.

The U.S. net worth data used in our study is the private sector's national wealth derived from two major data sources. Our data for the period 1896 through 1945 is based on Goldsmith's Table W1 estimates. We derived our private wealth series by subtracting the government and the public wealth from the national wealth in the original Table W1. This procedure was used to make this series comparable with another series that we had obtained for the subsequent 1946–1992 period providing only the private tangible assets. For this subsequent period, we used Table B11 of the Board of Governors of the Federal Reserve System, *Balance Sheets for the U.S. Economy 1945–1992*. Since this table gives a series for domestic private wealth, we added to this a series for U.S. net foreign assets from the same source to derive our national private wealth series for this period.

In the original tables of both Goldsmith et al. and the Federal Reserve Board, the asset figures are given at current cost, with the latter being the net of straight-line depreciation. The national wealth in these sources is in line with Kendrick's definition of wealth mentioned earlier, and consists of reproducible and nonreproducible tangible assets and net foreign assets. The reproducible assets include residential and nonresidential structures, consumer and producer durables, inventories, and monetary gold, silver, and SDRs, and the nonreproducible assets consist mainly of land.

B.2. Data Splicing for a Continuous Historical Series

To construct a continuous historical series of the U.S. GNP/wealth ratios, we spliced two series of GNP, one prewar and the other postwar, into one continuous series and did the same for wealth. For the splicing of wealth series, noting that the two separate wealth series do have an overlapping period of from 1945 to 1949, we looked at the gap between the two during this period and inflated the prewar series (by Goldsmith et al.) by 6% to give an upward lift. The fitting went reasonably well. As for the GNP series, we repeated the same steps and inflated the prewar series by 0.69% to match with the later series. Inflating at this percentage gave a perfect fit for 1929–1945 in which the two series overlap.

For 1896–1945, our GNP series was based on U.S. Bureau of the Census (1975, Series F1), while for 1946–1992, we used the original data from the Bureau of Economic Analysis (1992, 1993).

Chapter 9

ECONOMIC CONSERVATION LAWS AS INDICES OF CORPORATE PERFORMANCE

1. Introduction

In modern finance and accounting theory several different criteria are considered for measuring corporate performance. In addition, investors in stocks, for example, will use *P–E* ratios, etc. to judge how high or low a company's stock price is relative to its profitability, i.e. dividends. Corporate managers will analyze management conditions collectively by considering whether or not profitability per unit capital is at a satisfactory level, or by looking at trends in output per worker (labor productivity). In the final analysis, however, managers' greatest concern is whether earnings at their institutions are satisfactory in some sense. Or the yardstick for successful corporate management is whether corporate profits *are growing* at a satisfactory level.

Purely from the perspective of economics, a corporation may be analyzed on the assumption that it is operating to maximize profits over a certain period under the given conditions. In actuality, however, production costs and the exact workings of R&D, etc. at such an institution may be unpredictable. Since the numerical values of certain crucial parameters are ambiguous, it is impossible to judge whether a corporation's reported profits are being maximized or not. Under such circumstances it is not whether the company has achieved maximum profits but whether its income will continue to show *positive growth* that managers regard as the criterion of success. In short, there is a tendency to regard someone as a "good" manager as long as profits *are growing* even if he/she fails to achieve maximum profits under the given conditions.

This paper attempts to apply to corporate behavior at the microeconomic level an analytical method that has recently begun to gain the spotlight as a means of evaluating macroeconomic performance; namely, economic conservation laws Ramsey (1928). At the macro level, the efficient operation and performance of

The paper was presented at the 2003 Japan–US Technical Symposium on 4 April 2003, New York University. The author wishes to acknowledge helpful comments by Mariko Fujii, Gilbert Suzawa and Thomas Mitchell.

nations are analyzed on the assumption that they are carried out in ways that will maximize per capita consumption over the long-term. The theoretical tool derived from these results in the economic conservation law is known as the income to national wealth ratio. Here the criterion that performance is "good" because GDP *is rising* by a certain percent may not apply. Even if its GDP is increasing at a high growth rate, it is possible that a country may contain structural contradictions which will eventually cause its economic growth to stall. (An example of this is the postwar Japanese economy.) Likewise, a country with a low growth rate whose economy is forced into low growth because of international balance of payments deficits or budget deficits may still show reserves of strength as a comparatively well-balanced economy. (The United States is an example of this.) An analysis using economic conservation laws discerns the true long-term economic health of these two types of countries not from such criteria as growth rates or other such variable indicators of growth but from the invariance of certain variables. As a recent study in this area (Sato, 2002) shows, when the macro-management of the postwar US economy is compared with that of Japan from the standpoint of optimum theory, it is clear that the US is closer to the optimum.

Corporate management, too, needs to be able to judge whether or not under the given conditions a company is being managed in a way that will make long-term profits grow at their maximum. The aim of this paper is to apply the macroeconomic methods described above to microeconomic organizations and to come up with an assessment of corporate performance by applying economic conservation laws to individual companies and individual industries.

Because theoretical analyses using economic conservation laws are not yet in general use, the paper begins with a general model and a simple explanation of the Noether theorem for deriving conservation laws. Next, the forms that corporate profit maximization take often differ greatly depending on the nature of management and the management variables that are emphasized. In order to analyze these forms, several representative corporate behaviors found in the existing literature are presented as examples.

The next section focuses exclusively on the view of long-term profit maximization shared by these examples and tries to verify the workings of their profit maximizing behavior by using the minimum necessary data for an empirical analysis. As a theoretical assumption in such circumstances, the important criterion is whether or not the discount rate is fixed. Economic conservation laws under different hypotheses are then derived from this.

Data from the US automobile industry and aircraft manufacturing industry are considered as preliminary applications of these laws. The laws are also applied to data from representative Japanese real estate companies as well as to such corporations as Sony and Toyota. Japanese real estate companies are thought to have suffered the greatest losses from the collapse of Japan's economic bubble at the beginning of the 1990s. It will be interesting to see how they have been managed with a view of long-term maximization of profits in the 10 years since then. Sony and Toyota, on the other hand, are regarded as two of the most profitable corporations in Japan. Is that in fact the case? These questions will be explored in this paper.

2. General model and Noether's invariance principle

Let us consider a profit equation of the firm Π, which is twice continuously differentiable in each of its $2n+1$ arguments. We then have the firm's long-run profit,

$$J(x) = \int_a^b \Pi(t, x(t), \dot{x}(t))dt \tag{1}$$

where x is the vector of all quantities and prices, which the firm controls to maximize the long-run profit $J(x)$, $x(t) = (x^1(t), \ldots, x^n(t))$, and $\dot{x}(t) = (\dot{x}^1(t), \ldots, \dot{x}^n(t))$ for $t \in [a, b]$. The firm's problem is to maximize the long-run profit

$$J(x) = \int_a^b \Pi(t, x(t), \dot{x}(t))dt \to \max \tag{2}$$

subject to the appropriate initial conditions. The necessary condition for the optimal solution is that the Euler–Lagrange equations vanish:

$$E_i = \frac{\partial \Pi}{\partial x^i} - \frac{d}{dt}\left(\frac{\partial \Pi}{\partial \dot{x}^i}\right) = 0, \quad i = 1, \ldots, n \tag{3}$$

Consider a local Lie group transformation system given by:

$$\begin{aligned} \bar{t} &= \phi(t, x, \varepsilon), \ \varepsilon = (\varepsilon^1, \ldots, \varepsilon^r), \\ \bar{x}^i &= \psi^i(t, x, \varepsilon), \ i = 1, \ldots, n. \end{aligned} \tag{4}$$

where $\varepsilon = (\varepsilon^1, \ldots, \varepsilon^r)$ is vector of r real, independent essential parameters, which satisfies,

$$\begin{aligned} \phi(t, x, 0) &= t \\ \psi^i(t, x, 0) &= x^i, \ i = 1, \ldots, n \end{aligned} \tag{5}$$

Expanding the right-hand sides of (4) in Taylor series around $\varepsilon = 0$, we obtain the infinitesimal transformations (or generators) as:

$$\begin{aligned} \tau_k(t, x) &= \frac{\partial \phi}{\partial \varepsilon^k}(t, x, 0), \quad k = 1, \ldots, r \\ \xi_k^i(t, x) &= \frac{\partial \psi^i}{\partial \varepsilon^k}(t, x, 0), \quad i = 1, \ldots, n \end{aligned} \tag{6}$$

Using the conventional symbol of the infinitesimal transformations we write (6) as:

$$X_k = \tau_k \frac{\partial}{\partial t} + \xi_k^i \frac{\partial}{\partial x^i} + \left(\frac{d\xi_k^i}{dt} - \dot{x}^i \frac{d\tau_k}{dt}\right)\frac{\partial}{\partial \dot{x}^i} \tag{6'}$$

(Einstein summation convention on i, i.e. $\xi_k^i(\partial/\partial x^i) = \sum_{i=1}^n \xi_k^i(\partial/\partial x^i)$, etc).

The firm's long-run profit (Eq. (1)) is invariant up to a divergence term if,

$$J(\bar{x}) - J(x) = \int_{\bar{a}}^{\bar{b}} \Pi\left[\bar{t}, \bar{x}(\bar{t}), \frac{d\bar{x}(\bar{t})}{d\bar{t}}\right] d\bar{t} - \int_{a}^{b} \Pi\left[t, x(t), \frac{dx(t)}{dt}\right] dt$$
$$= \varepsilon^k \Phi_k(t, x(t)) + \mathrm{O}(\varepsilon)$$

where $\bar{t}$ and $\bar{x}$ are defined by (4).

The above is equivalent to

$$\Pi\left[\bar{t}, \bar{x}(\bar{t}), \frac{d\bar{x}(\bar{t})}{d\bar{t}}\right] \frac{d\bar{t}}{dt} - \Pi\left[t, x(t), \frac{dx(t)}{dt}\right] = \varepsilon^k \frac{d\Phi_k}{dt}(t, x(t)) + O(\varepsilon)$$

where $d\Phi_k/dt$ is the total derivative of Φ_k.

A necessary condition for Eq. (1) to be invariant under the r-parameter transformations (6) (invariant under nullity) is that Π and its derivatives satisfy the r identities,

$$X_k \Pi + \Pi \frac{d\tau_k}{dt} = \frac{d\Phi_k}{dt}, \quad k = 1, \ldots, r$$

where X_k is defined by (6′) (see Sato (1981, 1999), p. 244, Eq. (17)).

Noether's invariance principle (see Sato (1981, 1999), Chapter 7) states.

If the fundamental integral (1) is invariant under the r-parameter family of (Lie group) transformations (6), then there exist r conservation laws.

That is to say that if the long-run profit (1) is maximized, and it is invariant under the transformations (6), then r distinct quantities Ω_k are constant along the extremal, where

$$\Omega_k = -H\tau_k + \frac{\partial \Pi}{\partial \dot{x}^i} \xi_k^i - \Phi_k = \text{const}, \quad k = 1, \ldots, r \tag{7}$$

where H is the Hamiltonian defined by

$$H = -\Pi + \dot{x}^i \frac{\partial \Pi}{\partial \dot{x}^i} \tag{8}$$

and Φ_k is the term associated with the exact differential which is determined by (1) and (6). Since the addition of the total exact differentials $d\Phi_k/dt$ to (1) does not change the original Euler–Lagrange Eq. (3), Φ_k is called as the term associated with the dynamic invariance with nullity (see Sato (1981, 1999)).

3. Specialized Examples from the Literature

3.1. Capital (Physical or human) Investment Model (Kamien and Schwartz, 1981, p. 7)

Let $P(k)$ be the profit rate that is generated with a stock of physical capital k (or human capital). Assuming that the firm has little control over the price of the product, the revenue is $pf(k) = P(k)$, where p = price of the output = constant and $f(k)$ = output that k yields. Let $C(I)$ be the cost of gross investment, with the capital stock decaying at a constant proportionate rate b (or $C(I)$ = the cost of education or training with $b =$ the rate of forgetting for human capital). Then $C(I) = C(\dot{k} + bk)$. The firm's problem is to maximize the present value of the net profit stream over time, $\Pi = P(k) - C(\dot{k} + bk)$

$$\max \int_0^\infty e^{-\rho t}\Pi(k, \dot{k})\, dt = \max \int_0^\infty e^{-\rho t}[P(k) - C(\dot{k} + bk)]dt$$

subject to the initial conditions.

3.2. Price–quantity Adjustment Model

Consider a typical oligopolistic market. The profit function of a producer is dependent on p, the price of the product, q, the quantity of the product sold and $\dot{q}$, the rate of change of q:

$$\Pi(p,\ q,\ \dot{q}).$$

But, since other producers will also participate in the market, the market price will change depending upon the present level of p and the quantity supplied q. Thus we have

$$\dot{p} = f(p, q).$$

The firm's problem now is to maximize

$$\max \int_0^\infty e^{-\rho t}\Pi(p,\ q,\ \dot{q})dt$$

subject to

$$\dot{p} = f(p,\ q)$$

and the initial conditions.

The above will be reduced to

$$\max \int_0^\infty e^{-\rho t}(\Pi(p,\ q,\ \dot{q}) + \lambda(\dot{p} - f(p,\ q)))dt$$

subject to the appropriate initial conditions.

In general the price–quantity adjustment model assumes

$$\max \int_0^\infty e^{-\rho t}\Pi(\lambda,\ p,\ q,\ \dot{p},\ \dot{q})dt$$

subject to the initial conditions.

3.3. R&D Profit Maximization Model (Kamien and Schwartz, 1981; Sato and Suzawa, 1983; Sato, 2001; and Sato et al., 1999)

A typical firm in oligopolistic competition attempts to increase its profit by adjusting its output and by adopting cost-reducing innovations. The firm engages in research of both basic and applied types. Applied research which has a direct impact on reducing the cost of production of the output has to be supported by basic research; in fact, in many cases applied research can not be effective unless it is supported by basic research. In other words, basic research is an essential factor in the production of applied innovation.

Also, it should be noted that the R&D activities are dynamic in nature. The firm's problem is to maximize

$$J = \int_0^\infty e^{-\rho t}[R(p, y) - C(A, y) - S(A, \dot{A}, B) - T(B, \dot{B})]dt$$

subject to the intial conditions, where p is the price of the product, y the level of output, A the level of applied research, B the level of basic research, C the production cost, and S and T are R&D expenditures for applied and basic research.

3.4. No Stylized Facts

One purpose of this paper is to determine how useful the concept of conservation laws is in evaluating the actual performance of a corporation. This may be accomplished by observing the behavior of the corporation and analyzing the published information in its balance sheet and revenue and loss statement. The problem, however, is that one can not, a priori, determine to which category a particular company belongs. Thus, we must develop a general model not dependent on the particular aspects of the corporation's behavior exemplified in the above examples. Hence, we return to the general model to derive general conservation laws.

4. Conservation Laws as Indices of Corporate Performance

Returning to the general case of the firm's maximizing behavior (Eq. (1)), we write the invariance condition with nullity,

$$X\Pi + \Pi \frac{d\tau}{dt} = \frac{d\Phi}{dt}$$

as

$$\tau(t, x)\frac{\partial \Pi}{\partial t} + \xi^i(t, x)\frac{\partial \Pi}{\partial x^i} + \left(\frac{d\xi^i}{dt} - \dot{x}^i \frac{d\tau}{dt}\right)\frac{\partial \Pi}{\partial \dot{x}^i} + \Pi \frac{d\tau}{dt} = \frac{d\Phi}{dt} \tag{9}$$

(subscript $k = 1, \ldots, r$ is omitted in order to simplify the notation).

The left-hand side of the above is equal to

$$\frac{d}{dt}\left[\left(\Pi - \dot{x}^i \frac{\partial \Pi}{\partial \dot{x}^i}\right)\tau + \frac{\partial \Pi}{\partial \dot{x}^i}\xi^i\right] = \frac{d\Phi}{dt} \tag{10}$$

The above is known as *Noether's Theorem* (see Noether (1918), Logan (1997), Sato (1985, 1999)).

To prove this assertion, differentiate the left-hand side with respect to t, we get

$$\begin{aligned}
&\frac{\partial \Pi}{\partial t}\tau + \frac{\partial \Pi}{\partial \dot{x}^i}\left(\frac{d\xi^i}{dt} - \dot{x}^i\frac{d\tau}{dt}\right) + \Pi\frac{d\tau}{dt} + \dot{x}^i\tau\left[\frac{\partial \Pi}{\partial x^i} - \frac{d}{dt}\left(\frac{\partial \Pi}{\partial \dot{x}^i}\right)\right] + \frac{d}{dt}\left(\frac{\partial \Pi}{\partial \dot{x}^i}\right)\xi^i \\
&= \frac{\partial \Pi}{\partial t}\tau + \frac{\partial \Pi}{\partial x^i}\xi^i + \frac{\partial \Pi}{\partial \dot{x}^i}\left(\frac{d\xi^i}{dt} - \dot{x}^i\frac{d\tau}{dt}\right) + \Pi\frac{d\tau}{dt} + E_i(\dot{x}^i\tau - \xi^i) \\
&= X\Pi + \Pi\frac{d\tau}{dt} = \frac{d\Phi}{dt}
\end{aligned}$$

where E_i is the Euler–Lagrange Eq. (3) and $E_i = 0$. Hence we have,

$$\frac{d}{dt}\left[\left(\Pi - \dot{x}^i\frac{\partial \Pi}{\partial \dot{x}^i}\right)\tau + \frac{\partial \Pi}{\partial \dot{x}^i}\xi^i\right] = \frac{\partial \Pi}{\partial t}\tau + \frac{\partial \Pi}{\partial x^i}\xi^i + \frac{\partial \Pi}{\partial \dot{x}^i}\left(\frac{d\xi^i}{dt} - \dot{x}^i\frac{d\tau}{dt}\right) + \Pi\frac{d\tau}{dt} = \frac{d\Phi}{dt} \quad \textbf{(11)}$$

Hence

$$\left[\left(\Pi - \dot{x}^i\frac{\partial \Pi}{\partial \dot{x}^i}\right)\tau + \frac{\partial \Pi}{\partial \dot{x}^i}\xi^i\right] = -\int_t^\infty \left[\frac{\partial \Pi}{\partial s}\tau + \frac{\partial \Pi}{\partial x^i}\xi^i + \frac{\partial \Pi}{\partial \dot{x}^i}(\dot{\xi}^i - \dot{x}^i\dot{\tau}) + \Pi\dot{\tau}\right]ds \quad \textbf{(11')}$$

If we define $p^i = -\partial\Pi/\partial\dot{x}^i =$ supply price of x^i, the invariance condition (10) implies that the rate of change of the following quantity is always equal to zero:

$$\begin{aligned}
&(\text{profit} + \text{additional value created}) \times \text{time transformation} \\
&\quad - \sum_{i=1}^{n}(\text{supply price of} x^i) \times x^i\text{transformation} - \text{null term}\frac{d\Phi}{dt} = 0
\end{aligned}$$

There are several interesting cases where the infinitesimal transformations τ and ξ^i take special values and/or the profit function Π has a special form.

4.1. Conservation of the Value of the Firm

4.1.1. Time Transformation

Time transformation, i.e. $\tau \neq 0$ and $\xi^i = 0$, $i = 1, \ldots, n$,

From (11) we have

$$\begin{aligned}
\frac{d}{dt}[(\Pi + p^i\dot{x}^i)\tau] &= \tau\frac{\partial \Pi}{\partial t} + (\Pi + p^i\dot{x}^i)\frac{d\tau}{dt} \\
[(\Pi + p^i\dot{x}^i)\tau]_{s=t}^{b} &= \int_t^b \left[\tau\frac{\partial \Pi}{\partial s} + (\Pi + p^i\dot{x}^i)\frac{d\tau}{ds}\right]ds \\
[(\Pi + p^i\dot{x}^i)\tau]_{s=t} &= -\int_t^\infty \left[\tau\frac{\partial \Pi}{\partial s} + (\Pi + p^i\dot{x}^i)\frac{d\tau}{ds}\right]ds \qquad \textbf{(12)}
\end{aligned}$$

where, by the transversality condition, we assume

$$(\Pi + p^i \dot{x}^i)\tau \to 0 \text{ and } \left[\tau \frac{\partial \Pi}{\partial t} + (\Pi + p^i \dot{x}^i)\frac{d\tau}{dt}\right] \to 0 \text{ as } t \to \infty$$

The left-hand side of the above (12) represents the income-creation term (profit + change of the value of the firm) adjusted by the time transformation, while the right-hand side expresses income creation from the generalized value of the firm.

4.1.2. Constant Time Transformation

Constant time transformation, i.e. $\xi^i = 0$, $i = 1, \ldots, n$ and $\tau = 1$.
The above conservation law takes a more familiar form when $\tau = 1$. In this case,

$$(\Pi + p^i \dot{x}^i) = -\int_t^\infty \frac{\partial \Pi}{\partial s} ds = \text{income creation} \tag{13}$$

4.1.3. Constant Discount Rate

Constant discount rate, i.e. $\xi^i = 0$, $i = 1, \ldots, n$, $\tau = 1$ and

$$\Pi = D(t)G(x, \dot{x}) \tag{14}$$

We then have

$$(\Pi + p^i \dot{x}^i) = -\int_t^\infty \frac{dD(s)}{ds} G(x, \dot{x}) ds \tag{15}$$

If we assume that

$$D(t) = e^{-\rho t}, \quad \rho = \text{const} > 0 \tag{16}$$

The conservation law (15) has the familiar form

$$(\Pi + p^i \dot{x}^i) = \rho \int_t^\infty e^{-\rho s} G(x, \dot{x}) ds \tag{17}$$

or by redefining $p^i = e^{-\rho t}\pi i$, we have

$$(G + \pi^i \dot{x}^i) = \rho \int_t^\infty e^{-\rho(s-t)} G(x, \dot{x}) ds \tag{17'}$$

The current value of profit and changes in the value of the firm must be equal to the discounted value of the firm (see Weitzman (1976) and Sato (1981, 1999)).

We can give (17′) an alternative interpretation. By integrating Φ as:

$$\begin{aligned}
\Phi|_\tau^\infty = -\rho \int_t^\infty e^{-\rho s} G(s) ds \text{ or } \Phi|_\infty^\tau &= \rho \int_t^\infty e^{-\rho s} G(s) ds = \Phi(t) - \Phi(\infty) \\
&= \rho \left[\int_t^0 e^{-\rho s} G(s) ds - \int_\infty^0 e^{-\rho s} G(s) ds\right] \\
&= \rho \left[\int_0^\infty e^{-\rho s} G(s) ds - \int_0^t e^{-\rho s} G(s) ds\right] \\
e^{-\rho t}[G(t) + \pi^i(t)\dot{x}(t)] &= \rho \left[\int_0^\infty e^{-\rho s} G(s) ds - \int_0^t e^{-\rho s} G(s) ds\right]
\end{aligned} \tag{18}$$

The discounted total income creation is equal to the discount rate multiplied by the difference between the maximum discounted value of the firm and its discounted value up to the time period t.

4.1.4. Variable Discount Factor

Variable discount factor, i.e. $\xi^i = 0$, $i = 1, \ldots, n$, $\tau = 1$ and

$$\Pi(t) = e^{-\rho(t)}G(x(t), \dot{x}(t)) \tag{19}$$

We immediately obtain from (13)

$$(\Pi + p^i\dot{x}^i) = \int_t^{\infty} \rho'(s)e^{-\rho(s)}G(x(s), \dot{x}(s))ds \tag{20}$$

Alternatively from (11)

$$\frac{d}{dt}[(\Pi + p^i\dot{x}^i)] = \frac{\partial \Pi}{\partial t}$$

we obtain

$$\frac{d}{dt}[e^{-\rho(t)}(G + \pi^i\dot{x}^i)] = \rho'(t)\pi^i\dot{x}^i \tag{21}$$

where

$$\rho'(t) = \frac{d\rho}{dt} \tag{22}$$

Eq. (21) is the generalized conservation law when the discount rate is variable. When $\rho(t) = \rho t$, where ρ is a constant, we get the well-known Samuelson conservation law (Samuelson, 1970 and 1990).

4.1.5. Variable Discount Rate Combined with Variable Time Transformation

When τ, the infinitesimal transformation of time, takes a special form in (11′) together with $\xi^i = 0$, $i = 1, \ldots, n$, $\Pi = e^{-\rho(t)}G(x, \dot{x})$, and

$$\tau = \frac{1}{\rho'(t)} \tag{23}$$

We obtain a very interesting conservation law similar to the well-known types. From (11) we obtain

$$\frac{d}{dt}[(\Pi + p^i\dot{x}^i)\tau] = \frac{\partial \Pi}{\partial t}\tau + (\Pi + p^i\dot{x}^i)\frac{d\tau}{dt}$$

where

$$\Pi = e^{-\rho(t)}G(x, \dot{x}).$$

From

$$\frac{d}{dt}\left[\frac{e^{-\rho(t)}}{\rho'(t)}(G+\pi^i\dot{x}^i)\right] = -\rho'(t)e^{-\rho(t)}G\frac{1}{\rho'(t)} + e^{-\rho(t)}(G+\pi^i\dot{x}^i)\frac{d}{d}t\left(\frac{1}{\rho'(t)}\right)$$

$$= -e^{-\rho(t)}G - e^{-\rho(t)}(G+\pi^i\dot{x}^i)\frac{\rho''(t)}{(\rho'(t))^2}$$

$$= -e^{-\rho(t)}\left[G + (G+\pi^i\dot{x}^i)\frac{\rho''(t)}{(\rho'(t))^2}\right]$$

$$\int_t^{\infty}\frac{d}{ds}\left[\frac{e^{-\rho(t)}}{\rho'(s)}(G+\pi^i\dot{x}^i)\right]ds = -\int_t^{\infty} e^{-\rho(s)}\left[G+(G+\pi^i\dot{x}^i)\frac{\rho''(s)}{(\rho'(s))^2}\right]ds.$$

$$\left[\frac{e^{-\rho(s)}}{\rho'(s)}(G+\pi^i\dot{x}^i)\right]_{s=t}^{s=\infty} = -\int_t^{\infty} e^{-\rho(s)}\left[G+(G+\pi^i\dot{x}^i)\frac{\rho''(s)}{(\rho'(s))^2}\right]ds$$

As we assume that

$$\frac{e^{-\rho(t)}}{\rho'(t)}(G+\pi^i\dot{x}^i) \to 0, \quad \text{as} t \to \infty,$$

we get

$$[G+\pi^i\dot{x}^i]_{s=t} = \rho'(t)\int_t^{\infty} e^{-\rho(s)+\rho(t)}\left[G+(G+\pi^i\dot{x}^i)\frac{\rho''(s)}{(\rho'(s))^2}\right]ds \qquad \textbf{(24)}$$

or

$$\frac{[G+\pi^i\dot{x}^i]_{s=t}}{\int_t^{\infty} e^{-\rho(s)+\rho(t)}[G+(G+\pi^i\dot{x}^i)\rho''(s)/(\rho'(s))^2]ds} = \rho'(t), \qquad \textbf{(25)}$$

Thus the ratio of the current creation of the corporate income and the "adjusted" value of the corporation is equal to the rate of change in the discount factor at each period t.

Obviously when the discount rate is constant, i.e.

$$\rho(t) = \rho t$$

or

$$\rho'(t) = \rho = \text{const} > 0,$$

the above reduces the well-known conservation law expressed by (17′).

4.1.6. Conservation Laws When There Exist Time and Quantity (Price) Transformations

We shall briefly study the case where not only the time transformation τ but also the quantity (price) transformations ξ^i are not equal to zero, i.e. $\tau \neq 0$, $\xi^i \neq 0$. In particular, we will study the case,

$$\tau = 1 \quad \text{and} \quad \xi^i = x^i, \ = 1, \ldots, n \qquad \textbf{(26)}$$

This implies that the quantity transformations are subjected to the quantity augmenting transformations

$\bar{x}^i = Ax^i$, $\xi^i = x^i$, $i = 1, \ldots, n$, where A is a function of an essential parameter a, together with

$$\bar{t} = t + a \ (\tau = 1).$$

The conservation law is derived from (11)

$$\frac{d}{d}t[(\Pi + p^i \dot{x}^i) - p^i x^i] = \frac{d\Phi}{dt} = \frac{\partial \Pi}{\partial t} + \frac{\partial \Pi}{\partial x^i}\dot{x}^i - p^i \dot{x}^i$$

or

$$[\Pi + p^i \dot{x}^i - p^i x^i]_{s=t} = -\int_t^{\infty} \left(\frac{\partial \Pi}{\partial s} + \frac{\partial \Pi}{\partial x^i} x^i - p^i \dot{x}^i \right) ds \tag{27}$$

If we simplify Π as $\Pi = e^{-\rho t} G(x, \dot{x})$, the above reduces to

$$[G + \pi^i \dot{x}^i - \pi^i x^i]_t = \rho \int_t^{\infty} e^{-\rho(s-t)} [G + r^i x^i - \pi^i x^i] ds \tag{28}$$

where $r^i = \partial G / \partial x^i$ is the marginal profit of x^i.

The current income creation subtracted by $\pi^i x^i$ will be always equal to the "adjusted" value of the company (adjustments being made by $r^i x^i$ and $\pi^i x^i$) multiplied by the discount rate ρ.

5. Actual Corporate Performance and Conservation Laws: Preliminary Investigation

If a firm engages in long-run profit maximizing operations in whatever form it may be, it is conceivable that we may be able to observe the outcome of this behavior from the published market data. In this section we attempt to discover if companies perform as the theoretical model prescribes. In other words, we test various types of conservation laws against the data from several industries and companies in the US and in Japan. Obviously even if we cannot find "good" results, it does not mean that the companies are not adhering to profit maximizing behavior and therefore their corporate performances are not good. It simply may suggest that the model itself is too simple or that the firm must have been forced to change the maximizing behavior as the surrounding conditions have changed.

5.1. US Data

To show the result of an empirical application (see Lian (1995)), we look at the data relating with two sectors of the US economy; the automobile industry and

TABLE 1. Test of Micro-conservation Law.

Year	Profit	Market value	ρ	Profit MA (5)	Market value (5)	ρ (5)
1976	235.30	1962.59				
1977	335.66	2323.20	0.300			
1978	505.86	4415.03	0.588			
1979	724.07	4810.29	0.233			
1980	775.77	6425.00	0.372	515.33	3987.22	
1981	670.09	3604.90	−0.596	602.29	4315.68	0.216
1982	596.99	5458.10	0.449	654.56	4942.66	0.259
1983	740.64	7263.09	0.350	701.51	5512.28	0.231
1984	1220.72	9173.60	0.341	800.84	6384.94	0.262
1985	993.24	12024.16	0.320	844.34	7504.77	0.262
1986	1021.19	11618.50	0.053	914.56	9107.49	0.276
1987	820.05	8501.17	−0.270	959.17	9716.10	0.161
1988	1050.47	12810.54	0.418	1021.13	10825.59	0.197
1989	705.26	16542.01	0.268	918.04	12299.28	0.194
1990	1745.57	17722.24	0.165	1068.51	13438.89	0.164
1991	2021.34	19691.44	0.203	1268.54	15053.48	0.192
1992	2371.86	16331.84	−0.060	1578.90	16619.61	0.189

the aircraft industry (1976–1992). From the automobile industry, we chose GM, Ford and Chrysler. We assumed that these companies are working under the condition of the constant discount rate, i.e. Eq. (17′). The ratio of the income creation to the value of the company was not invariant. Even using the moving average data the result was not constant as the model predicts.

From the aircraft industry, we combined the data of Boeing Co., Grumman Co., and McDonnell Douglas Co. The data and results of calculation are shown in Table 1. The ratios by using the original data are unstable (see

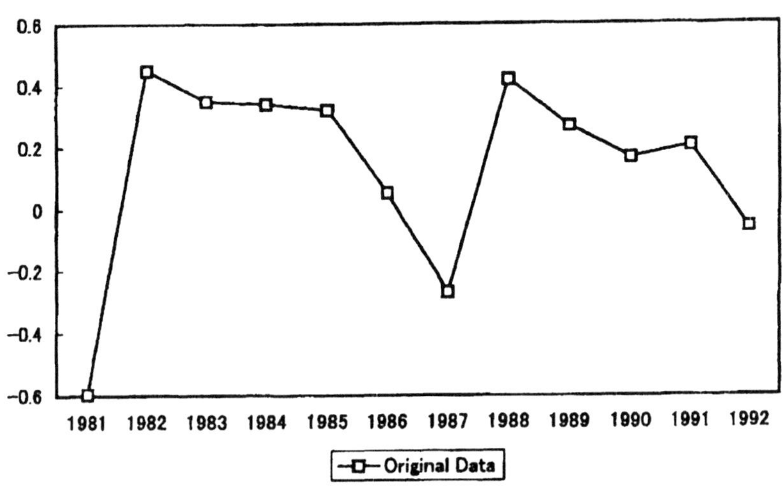

FIGURE 1. Micro-conservation Law (US Aircraft Industry).

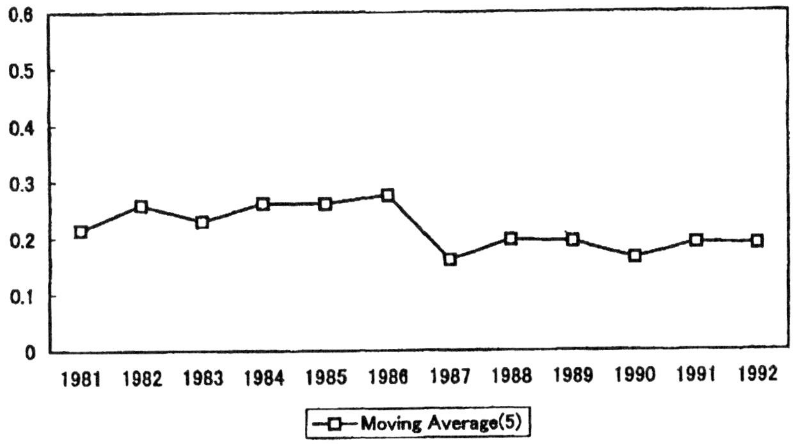

FIGURE 2. Micro-conservation Law (US Aircraft Industry).

Fig. 1), but the ratios by using the moving average data are quite stable (see Fig. 2).

For the moving average data of the aircraft industry, the average ratio for the whole period examined is 0.217. However, there is a shift of pattern for this figure. From 1981 to 1986, the average is 0.251. The average for the period of 1987–1992 is 0.183.

The results were as expected and shows that the automobile industry in the United States has been getting more and more competitive. The aircraft industry, however, will probably remain a monopolistic industry.

5.2. Japanese Data

This method was applied to five companies to test their performance for the period 1994–2002. These companies are Toyota and Sony, which are considered to be most profitable Japanese companies even during this period of the stagnant economy, and three real estate companies (Fudosan Kaisha), Mitsui, Mitsubishi and Sumitomo. These three companies represent the business sector which enjoyed enormous abnormal profits during the bubble period, but suffered losses resulting from the subsequent persistent stagnation.

A preliminary examination reveals that there are no clear trends as far as the ratios are concerned. This is to be expected since the Japanese companies basically tried to survive by making a short-run decision during this turbulent post-bubble period. We plan to attempt to extend the analysis to cover the pre-bubble and bubble periods.

Tables 2 and 3 and Fig. 3 show the basic data and the test of the conservation law (Eq. (17′))

TABLE 2. Test of Micro-conservation Law (Japanese Companies, in Hundred Million Yen).

	March 1993	March 1994	March 1995	March 1996	March 1997	March 1998	March 1999	March 2000	March 2001	March 2002
MitsuiFudosan Co. Ltd.										
Current revenue	1108	835	760	676	−14	636	569	819	1083	1029
Total value at the end of fiscal year	10061	9374	7764	11292	10401	10319	8670	7963	9808	8475
Rate of return (percent)		1.6	−10.9	37.2	−8.7	5.4	−12.5	1.4	29.9	−3.6
Current revenue MA (3)			901.0	757.0	474.0	432.7	397.0	674.7	823.7	977.0
Total value at the end of fiscal year MA (3)			9066.3	9476.7	9819.0	10670.7	9796.7	8984.0	8813.7	8748.7
Rate of return MA (3) (percent)				12.3	8.3	12.0	−4.9	−1.5	7.4	10.4
Mitsubishi Co. Ltd.										
Current revenue	1113	972	915	816	624	583	688	682	806	769
Total value at the end of fiscal year	13382	13512	12693	19098	17149	16889	15655	14460	15071	12108
Rate of return (percent)		8.2	0.8	37.8	−7.7	1.9	−3.5	−3.5	9.4	−18.1
Current revenue MA (3)			1000.00	901.00	785.00	674.33	631.67	651.00	725.33	752.33
Total value at the end of fiscal year MA (3)			13195.67	15101.00	16313.33	17712.00	16564.33	15668.00	15062.00	13879.67
Rate of return MA (3) (percent)				18.6	12.2	11.7	−3.1	−1.6	0.8	3.1
Sumitomo Realty & Development Co. Ltd.										
Current revenue	593	538	457	509	433	449	565	705	752	804
Total value at the end of fiscal year	2835	2685	2278	3340	3385	3198	1827	1412	2457	2723
Rate of return (percent)		14.5	2.2	47.0	14.1	8.2	−44.1	20.5	73.1	39.3
Current revenue MA (3)			529.33	501.33	466.33	463.67	482.33	573.00	674.00	753.67
Total value at the end of fiscal year MA (3)			2599.33	2767.67	3001.00	3307.67	2803.33	2145.67	1989.67	2197.33
Rate of return MA (3) (percent)				24.2	23.3	23.3	−0.8	−3.9	22.5	47.9

Sony Corporation									
Current revenue		−1666	2353	3525	5141	3381	2232	2253	1346
Total value at the end of fiscal year	21874	16265	23903	33232	46013	44943	65778	81846	61623
Rate of return (percent)		−44.7	41.8	38.7	38.9	5.1	35.1	22.4	−30.6
Current revenue MA (3)				1404.00	3673.00	4015.67	3584.67	2622.00	1943.67
Total value at the end of fiscal year MA (3)			20680.67	24466.67	34382.67	41396.00	52244.67	64189.00	69749.00
Rate of return MA (3) (percent)				21.2	39.5	26.6	27.6	22.7	10.8
Toyota Motor Corporation									
Current revenue		3410	3481	6651	8812	7749	7760	8701	11235
Total value at the end of fiscal year	73639	66163	88547	118658	135.042	128.934	198.207	158.166	129.794
Rate of return (percent)		−6.1	29.2	31.0	18.2	1.3	38.9	−19.8	−13.2
Current revenue MA (3)				4514.00	6104.67	7527.33	7897.00	8070.00	9232.00
Total value at the end of fiscal year MA (3)			76116.33	91122.67	114082.33	127544.67	154061.00	161769.00	162055.67
Rate of return MA (3) (percent)				21.4	25.5	16.5	22.3	9.8	5.9

TABLE 3. Moving Average

	March 1993	March 1994	March 1995	March 1996	March 1997	March 1998	March 1999	March 2000	March 2001	March 2002
MitsuiFudosan Co. Ltd.										
Current revenue	1108	835	760	676	−14	636	569	819	1083	1029
Total value at the end of fiscal year	10061	9374	7764	11292	10401	10319	8670	7963	9808	8475
Rate of return (percent)		1.6	−10.9	37.2	−8.7	5.4	−12.5	1.4	29.9	−3.6
Rate of return MA (3) (percent)				9.3	5.9	11.3	−5.3	−1.9	6.3	9.2
Mitsubishi Co. Ltd.										
Current revenue	1113	972	915	816	624	583	688	682	806	769
Total value at the end of fiscal year	13382	13512	12693	19098	17149	16889	15655	14460	15071	12108
Rate of return (percent)		8.2	0.8	37.8	−7.7	1.9	−3.5	−3.5	9.4	−18.1
Rate of return MA (3) (percent)				15.6	10.3	10.7	−3.1	−1.7	0.8	−4.1
Sumitomo Realty & Development Co. Ltd.										
Current revenue	593	538	457	509	433	449	565	705	752	804
Total value at the end of fiscal year	2835	2685	2278	3340	3385	3198	1827	1412	2457	2723
Rate of return (percent)		14.5	2.2	47.0	14.1	8.2	−44.1	20.5	73.1	39.3
Rate of return MA (3) (percent)				21.2	21.1	23.1	−7.3	−5.1	16.5	44.3
Sony Corporation										
Current revenue			−1666	2353	3525	5141	3381	2232	2253	1346
Total value at the end of fiscal year		21874	16265	23903	33232	46013	44943	65778	81846	61623

Rate of return (percent)		−44.7	41.8	38.7	38.9	5.1	35.1	22.4	−30.6
Rate of return MA (3) (percent)				11.9	39.8	27.6	26.4	20.9	8.9
Toyota Motor Corporation									
Current revenue		3410	3481	6651	8182	7749	7760	8701	11235
Total value at the end of fiscal year	73639	66163	88547	118658	135042	128934	198207	158166	129794
Rate of return (percent)		−6.1	29.2	31.0	18.2	1.3	38.9	−19.8	−13.2
Rate of return MA (3) (percent)				18.0	26.1	16.8	19.4	6.8	1.9

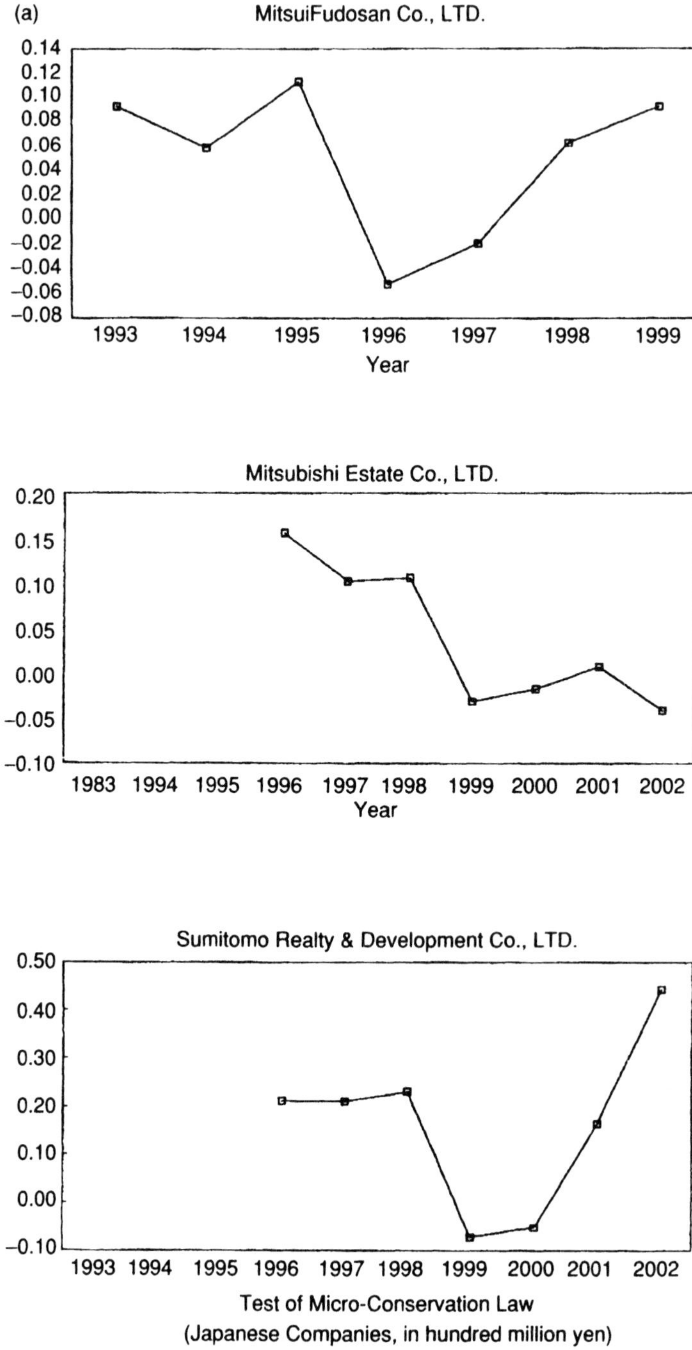

FIGURE 3. Test of Micro-conservation Law (Japanese Companies).

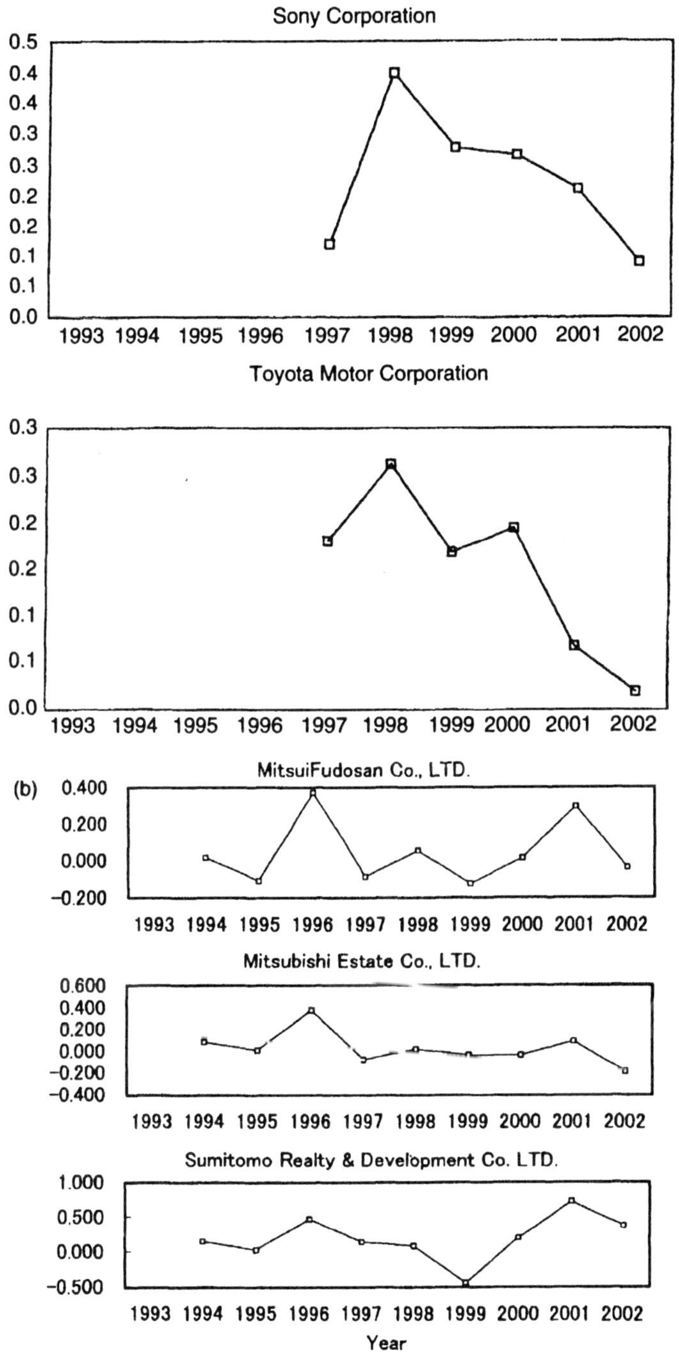

Figure 3. *(Continued)*

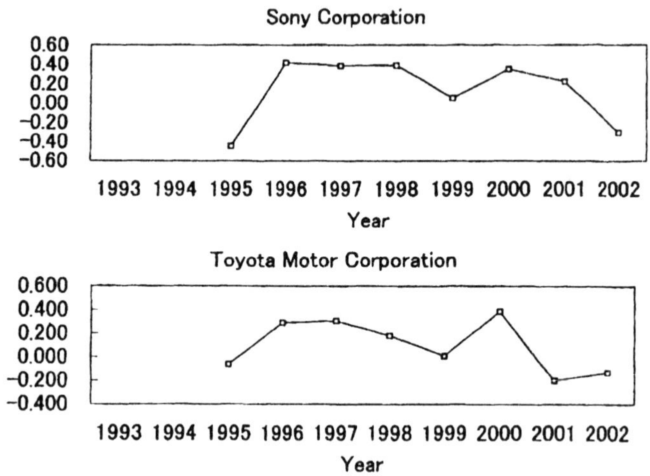

FIGURE 3. *(Continued)*

5.3. Remarks on Japanese Macro Trend

In an earlier paper (Sato, 2002), we discovered that, unlike the cases of the US and other advanced OECD countries, the Japanese economic performance seemingly did not follow the optimizing behavior under the constant discount rate (see Figs. 4 and 5).

Compared with the US case which is rather stable around 0.3, the Japanese economy exhibited the declining trend, which may require a different approach, i.e. the model of variable discount rate as discussed in the previous section (Eq. (25)). The simplest case which may fit to this trend is:

$$\rho'(t) = a - bt$$

or

$$\rho(t) = c + at - \frac{1}{2}bt^2, \quad a, b, c > 0 \tag{29}$$

It is well known that immediately after World War II, Japanese business executives had a rather myopic view on the planning of their business activity, but as the economy became more stable their behavior changed to adopting a longer time horizon. This might explain why the discount rate has gradually declined from 25 percent in 1955 to around 13 percent right before the bursting of the bubble period. During the bubble period the Japanese attitude was that the longer they wait the more money they can make. This is the period when Sony's President Morita insulted Americans with the statement that: "Americans try to make money in ten minutes, while we Japanese try in ten years".

It is, however, not easy to apply the model of "variable" discount rate for estimation purposes. We must, a priori, know the behavior of the discount rate factor such as a, b, c in Eq. (29) to estimate the "wealth-like" quantity in Eq. (25). Alternatively, we may want to apply Eq. (21) instead, by estimating the changes in

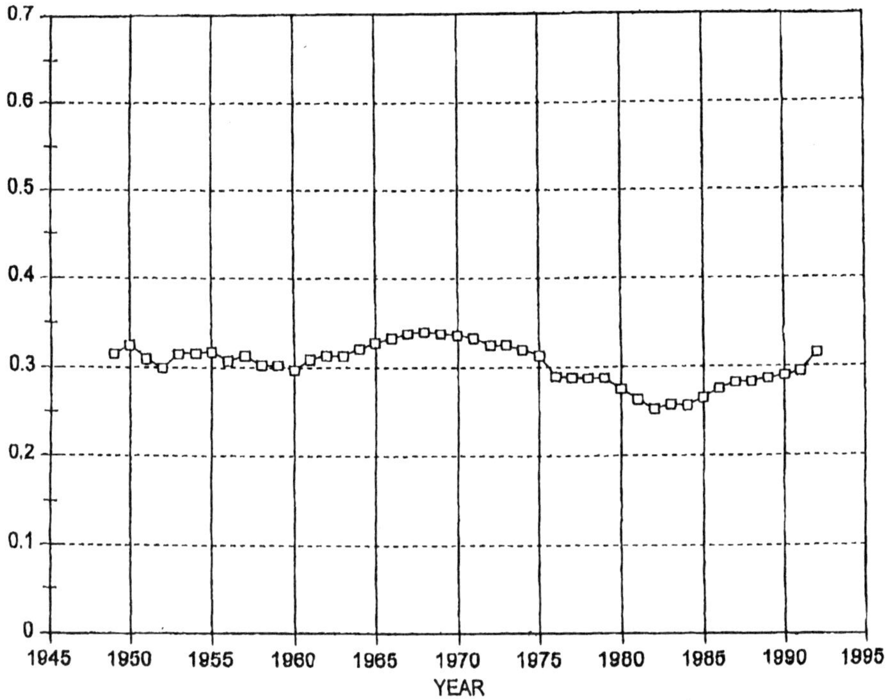

FIGURE 4. Income/wealth Ratio (Postwar).

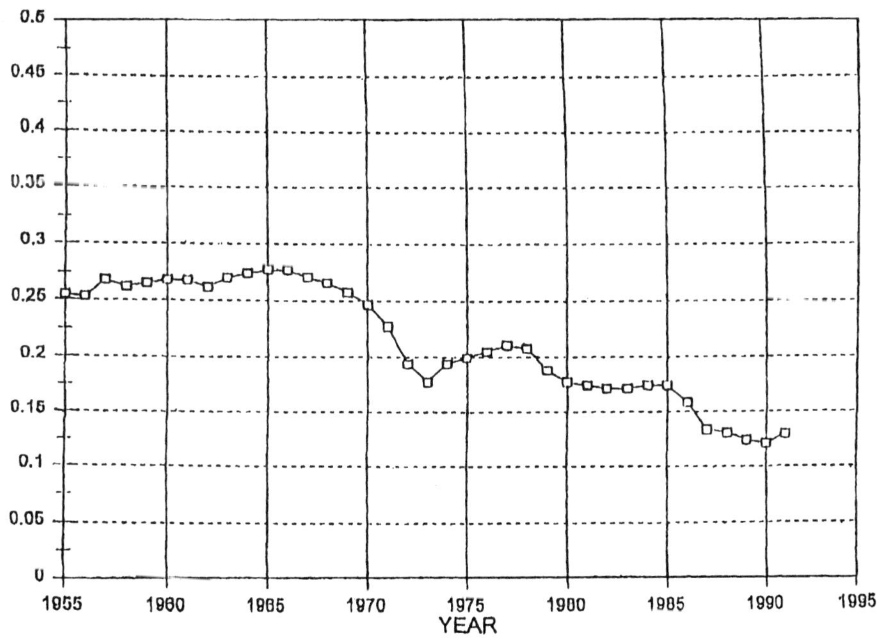

FIGURE 5. Ratio of GDP/net Wealth in Japan.

the national income and the value of investment. It should be noted that the three models, Eqs. (17′), (20) and (24) are identical if and only if $\rho'(t)$ is constant.

6. Concluding Remarks

This paper has attempted to analyze and apply the theory of economic conservation laws as a criterion for measuring corporate performance on a microeconomic level. Although it is not particularly our intention to make a close computational analysis of empirical results, we think that conservation laws ought to be used along with and as a supplement to various other indicators of corporate performance that have often traditionally been used in the areas of finance, accounting and management. The traditional standards of measurement found in management, accounting and finance theory remain as effective as ever, of course, but the new tool presented in this paper, we believe, can also play an effective role in measuring the workings of what theoretical economics regards as the ultimate goal of corporate performance, the maximization of profits.

References

Kamien, M., Schwartz, N., 1981. Dynamic Optimization. North-Holland, New York.

Lian, C., 1995. Optimal growth, technical progress and economic conservation laws. Ph.D. Dissertation, New York University, New York.

Logan, J.D., 1977. Invariant variational principles. In: Mathematics in Science and Engineering, vol. 138. Academic Press, New York.

Noether, E., 1918. Invariante Variationsprobleme, Nachr. Akad. Wiss. Göttingen. Math-Phys. K1 (II), 235–257. Invariant variation problems. Transport Theory and Statistical Physics 1, 186–207 (Tavel, M.A., Trans., 1971).

Ramsey, F., 1928. A mathematical theory of saving. Economic Journal 38, 543–559.

Samuelson, P., 1970. Law of conservation of the capital–output ratio. In: Sato, R., Ramachandran, R. (Eds.), Conservation Laws and Symmetry: Application to Economics and Finance. Kluwer Academic Publishers, Boston, 1990 (Chapter 2). Proceedings of the National Academy of Sciences, Applied Mathematical Sciences 67, 1477–1479 (reprint).

Samuelson, P., 1990. In: Sato, R., Ramachandran, R. (Eds.), Two Conservation Laws in Theoretical Economics in Conservation Laws and Symmetry. Kluwer Academic Publishers, Boston (Chapter 3).

Sato, R., 1981. Theory of Technical Change and Economic Invariance: Application of Lie Groups. Academic Press, New York (revised reprint version, E. Elgar, 1999).

Sato, R., 1985. The invariance principle and income-wealth conservation laws: application of lie groups and related transformations. Journal of Econometrics 30, 365–389. Also in Production, Stability and Dynamic Symmetry: Selected Essays of Ryuzo Sato, vol. 2. Economists of the Twentieth Century Series, E. Elgar, 1999.

Sato, R., 2001. International competition and asymmetric technology game. Japan and the World Economy 13, 217–233.

Sato, R., 2002. Optimal economic growth: test of income–wealth conservation laws in OECD countries. Macroeconomic Dynamics 6, 548–572.

Sato, R., Suzawa, G., 1983. Research and Productivity: Endogenous Technical Change. Auburn House, Boston.

Sato, Ramachandran, R., Lian, C., 1999. A model of optimal economic growth and endogenous bias. Macroeconomic Dynamics 3.

Weitzman, M.L., 1976. On the welfare significance of national product in a dynamic economy. Quarterly Journal of Economics 90, 156–162.

Discussion

CONSERVATION LAWS FOR MICROECONOMISTS! COMMENTS ON "ECONOMIC CONSERVATION LAWS AS INDICES OF CORPORATE PERFORMANCE" BY RYUZO SATO

THOMAS MITCHELL*

Department of Economics, Southern Illinois University Carbondale, MC 4515, Carbondale, IL 62901-4515, USA

Received 4 April 2003; received in revised form 1 December 2003; accepted 8 December 2003

Finally! A class of economic conservation laws that even a microeconomic theorist can love! At the same time, however, some of us should be wondering why this did not occur to us earlier. Emmy Noether's theorem on conservation laws has been around for more than 80 years (Noether, 1918); Frank Ramsey's rule for an optimal saving policy has been around for more than 70 years (Ramsey, 1928); Paul Samuelson's paper that formally introduced conservation laws into economics has been around for more than 30 years (Samuelson, 1970); and Ryuzo Sato's book on Lie groups has been around for more than 20 years (Sato, 1981, 1999). Not surprisingly though, it is Sato who has assembled the pieces that are combined in the preceding paper.

While the mathematical ideas have been lying around for some time, and they have been applied previously to macroeconomic models, Sato's current paper is the first application of the concepts and mathematical techniques to a microeconomic environment. The paper creates vast opportunities for both theoretical and empirical work that offer the possibility for us to better analyze and understand modern corporate and industrial behavior.

1. A Different View of An Old Problem

Let $x\mathbb{R}^n$ denote a vector of quantities, prices and other relevant [state] variables. Sato forms the long-run profit-maximization problem as a variational problem,

*Tel.: +1-618-453-5073.
E-mail address: tmitch@siu.edu (T. Mitchell).

0922-1425/$ – see front matter doi:10.1016/j.japwor.2003.12.004

$$\max J[\boldsymbol{x}] = \int_a^b \Pi(t, \boldsymbol{x}(t), \boldsymbol{x}(t))dt. \tag{1}$$

The associated Euler equation are:

$$\frac{\partial \Pi}{\partial x_i} - \frac{d}{dt}\left(\frac{\partial \Pi}{\partial \dot{x}_i}\right) = 0, \qquad i = 1, \ 2, \ \ldots, n.$$

We add to the problem a Lie group of transformations, acting on some or all of the economic variables, $\boldsymbol{x}$, and time, t:

$$\bar{x} = \boldsymbol{\psi}(\boldsymbol{x}, \ t, \ \boldsymbol{\varepsilon}), \tag{2}$$

$$\bar{t} = \phi(\boldsymbol{x}, \ t, \ \boldsymbol{\varepsilon}). \tag{3}$$

The vector-valued function $\boldsymbol{\psi}$ and the scalar-valued function ϕ depend on the values of $x_1, \ x_2, \ \ldots, \ x_n$, time ($t$), and r parameters, $\varepsilon_1, \ \varepsilon_2, \ \ldots, \ \varepsilon_r$. Further, the group properties are satisfied.

There are four group properties.

1. *There is a composite transformation.* For every pair of admissible values of the parameter vector defining two applications of the transformation, say $\boldsymbol{\varepsilon}'$ and $\boldsymbol{\varepsilon}''$, there is an admissible value $\hat{\varepsilon}$ such that

$$\bar{\bar{x}} = \boldsymbol{\psi}(\bar{x}, \ \bar{t}, \ \boldsymbol{\varepsilon}'') = \boldsymbol{\psi}[\boldsymbol{\psi}(\boldsymbol{x}, \ t, \ \boldsymbol{\varepsilon}'), \ \phi(\boldsymbol{x}, \ t, \ \boldsymbol{\varepsilon}'), \ \boldsymbol{\varepsilon}''] = \boldsymbol{\psi}(\boldsymbol{x}, \ t, \ \hat{\boldsymbol{\varepsilon}});$$
$$\bar{\bar{t}} = \phi(\bar{x}, \ \bar{t}, \ \boldsymbol{\varepsilon}'') = \phi[\boldsymbol{\psi}(\boldsymbol{x}, \ t, \ \boldsymbol{\varepsilon}'), \ \phi(\boldsymbol{x}, \ t, \ \boldsymbol{\varepsilon}'), \ \boldsymbol{\varepsilon}''] = \phi(\boldsymbol{x}, \ t, \ \hat{\boldsymbol{\varepsilon}}).$$

   In other words, for every pair of admissible transformations, there is a single admissible transformation that is equivalent.

2. *There is an identity transformation.* For our purposes, there exists a value $\boldsymbol{\varepsilon}_0 \in \mathbb{R}^r$, such that no change occurs:[1]

$$\boldsymbol{\psi}(\boldsymbol{x}, \ t, \ \boldsymbol{\varepsilon}_0) = \boldsymbol{x};$$
$$\phi(\boldsymbol{x}, \ t, \ \boldsymbol{\varepsilon}_0) = t \cdot$$

3. *There is an inverse transformation.* For our purposes, for every admissible $\boldsymbol{\varepsilon} \in \mathbb{R}^r$ there exists another admissible value $\gamma \in \mathbb{R}^r$, such that any change represented by $\boldsymbol{\varepsilon}$ can be undone, so that the "identity values" are restored:

$$\boldsymbol{\psi}(\bar{x}, \ \bar{t}, \ \gamma) = \boldsymbol{\psi}[\boldsymbol{\psi}(\boldsymbol{x}, \ t, \ \boldsymbol{\varepsilon}), \ \boldsymbol{\phi}(\boldsymbol{x}, \ t, \ \boldsymbol{\varepsilon}), \ \gamma] = \boldsymbol{x};$$
$$\boldsymbol{\phi}(\bar{x}, \ \bar{t}, \ \gamma) = \boldsymbol{\phi}[\boldsymbol{\psi}(\boldsymbol{x}, \ t, \ \boldsymbol{\varepsilon}), \ \boldsymbol{\phi}(\boldsymbol{x}, \ t, \ \boldsymbol{\varepsilon}), \ \gamma] = t.$$

4. *The associative law holds.* Two actions of the group followed by a single action of the group are equivalent to a single action followed by two actions that

[1] Sato takes $\boldsymbol{\varepsilon}_0 = 0$. There is no harm in doing so; indeed, the notation is probably simplified. However, $\boldsymbol{\varepsilon}_0$ need not be an r-element null vector.

have the same values of $\boldsymbol{x}$, t, $\boldsymbol{\varepsilon}^1$, $\boldsymbol{\varepsilon}^2$, and $\boldsymbol{\varepsilon}^3$. To understand the associative law, define functions $\boldsymbol{\Psi}$ and $\boldsymbol{\Phi}$ to represent two actions of the group:

$$\bar{\bar{x}} = \psi(\bar{x}, \bar{t}, \boldsymbol{\varepsilon}^2) = \psi[\psi(\boldsymbol{x}, t, \boldsymbol{\varepsilon}^1), \phi(\boldsymbol{x}, t, \boldsymbol{\varepsilon}^1), \boldsymbol{\varepsilon}^2] = \boldsymbol{\Psi}(\boldsymbol{x}, t; \boldsymbol{\varepsilon}^1, \boldsymbol{\varepsilon}^2);$$
$$\bar{\bar{t}} = \phi(\bar{x}, \bar{t}, \boldsymbol{\varepsilon}^2) = \phi[\psi(\boldsymbol{x}, t, \boldsymbol{\varepsilon}^1), \phi(\boldsymbol{x}, t, \boldsymbol{\varepsilon}^1), \boldsymbol{\varepsilon}^2] = \boldsymbol{\Phi}(\boldsymbol{x}, t; \boldsymbol{\varepsilon}^1, \boldsymbol{\varepsilon}^2)\cdot$$

With this notation, we can represent the associative law as:

$$\psi[\boldsymbol{\Psi}(\boldsymbol{x}, t; \boldsymbol{\varepsilon}^1, \boldsymbol{\varepsilon}^2), \boldsymbol{\Phi}(\boldsymbol{x}, t; \boldsymbol{\varepsilon}^1, \boldsymbol{\varepsilon}^2); \boldsymbol{\varepsilon}^3] = \boldsymbol{\Psi}[\psi(\boldsymbol{x}, t; \boldsymbol{\varepsilon}^1), \phi(\boldsymbol{x}, t; \boldsymbol{\varepsilon}^1); \boldsymbol{\varepsilon}^2, \boldsymbol{\varepsilon}^3];$$
$$\phi[\boldsymbol{\Psi}(\boldsymbol{x}, t; \boldsymbol{\varepsilon}^1, \boldsymbol{\varepsilon}^2), \boldsymbol{\Phi}(\boldsymbol{x}, t; \boldsymbol{\varepsilon}^1, \boldsymbol{\varepsilon}^2); \boldsymbol{\varepsilon}^3] = \boldsymbol{\Phi}[\psi(\boldsymbol{x}, t; \boldsymbol{\varepsilon}^1), \phi(\boldsymbol{x}, t; \boldsymbol{\varepsilon}^1); \boldsymbol{\varepsilon}^2, \boldsymbol{\varepsilon}^3].$$

Although it is not important for our purposes here, the group properties are not all independent. In particular, Mitchell and Primont (1991) demonstrate that the identity property may be deduced from the others.

The infinitesimal transformations of the group are obtained as the first-order terms of a Taylor series expansion, expanded around the "identity value" of the r-dimensional parameter vector, $\boldsymbol{\varepsilon}_0 \in \mathbb{R}^r$:

$$\xi_i^k(\boldsymbol{x}, t) = \frac{\partial \psi_i(\boldsymbol{x}, t, \boldsymbol{\varepsilon}_0)}{\partial \boldsymbol{\varepsilon}_k}, \quad i = 1, 2, \ldots, n; \; k = 1, 2, \ldots, r; \tag{4}$$

$$\tau_k(\boldsymbol{x}, t) = \frac{\partial \phi(\boldsymbol{x}, t, \boldsymbol{\varepsilon}_0)}{\partial \boldsymbol{\varepsilon}_k}, \quad k = 1, 2, \ldots, r. \tag{5}$$

The infinitesimal transformations in (4) and (5) are used to form the following, "infinitesimal" operator,

$$X_k = \tau_k \frac{\partial}{\partial t} + \sum_{i=1}^{n} \xi_i^k \frac{\partial}{\partial x_i} + \left(\frac{d\xi_i^k}{dt} - \dot{x}_i \frac{d\tau_k}{dt}\right) \frac{\partial}{\partial \dot{x}_i}. \tag{6}$$

The Noether theorem concerns the invariance of the variational problem in Eq. (1) even though the values of the parameters, $\varepsilon_1, \varepsilon_2, \ldots, \varepsilon_r$, are changing the values of x_i's and t:

If the variational problem (1) is invariant under the r-parameter group described by Eqs. (2) and (3),[2] then there exist r conservation laws: along any extremal there are r distinct quantities, $\Omega_1, \Omega_2, \ldots, \Omega_r$, which are constant if $J[x]$ is maximized and invariant under the operation of the group of transformations.

This is truly remarkable! From a theoretical perspective, the terminology *should* remind those who have studied physics of various conservation laws in that physical science: for example, the conservation of momentum, the conservation of angular momentum, the conservation of energy, and the conservation of electric charge.

The notion of "conservation" is the principle that some quantity within a system is unchanged, even though various other quantities within the system are changing. That the values of some variables are constant along an extremal for the problem in Eq. (1) presents fascinating possibilities on both a theoretical level

[2] The r-parameter group can also be defined through the infinitesimal transformations in Eqs. (4) and (5), or even through its operator(s), defined in Eq. (6).

and an empirical level. In the following sections we will use Sato's examples of microeconomic conservation laws for both review of his results and as opportunities to review some of the properties of Lie groups of transformations.

2. Examples of Microeconomic Conservation Laws

2.1. Conservation of the Value of the Firm

2.1.1. Time Transformation

The Lie group of transformations is specified as:

$$\bar{\boldsymbol{x}} = \psi(\boldsymbol{x},\ t,\ \mathbf{1}) \equiv \boldsymbol{x};$$
$$\bar{t} = \phi(\boldsymbol{x},\ t,\ \boldsymbol{\varepsilon})\cdot$$

That is, there is no group effect on the state variables. The infinitesimal transformations of the group are:

$$\xi_i^k(\boldsymbol{x},\ t) = \frac{\partial\psi_i(\boldsymbol{x},\ t,\ \boldsymbol{\varepsilon}_0)}{\partial\varepsilon_k} \equiv 0, \qquad i = 1,\ 2,\ \ldots,\ n;\ k = 1,\ 2,\ \ldots,\ r;$$
$$\tau_k(\boldsymbol{x},\ t) = \frac{\partial\phi(\boldsymbol{x},\ t,\ \boldsymbol{\varepsilon}_0)}{\partial\varepsilon_k}, \qquad k = 1,\ 2,\ \ldots,\ r,$$

so that the infinitesimal operator is given by:

$$X_k = \tau_k\frac{\partial}{\partial t} - \sum_{i=1}^{n}\dot{x}_i\frac{d\tau_k}{dt}\frac{\partial}{\partial\dot{x}_i}; \quad k = 1,\ 2,\ \ldots,\ r.$$

Suppose that the group operates so that the "transformation of time" is a simple shift with a single real parameter $\varepsilon \in \mathbb{R}$, $\bar{t} = \phi(\boldsymbol{x},\ t,\ \varepsilon) = \varepsilon + t$, and there is no effect on $\boldsymbol{x}$: $\psi(\boldsymbol{x},\ t,\ \varepsilon) \equiv \boldsymbol{x}$. In this case, the group action is described by a one-parameter group (i.e., $r = 1$) with the following properties.

1. The composition property is illustrated by:

$$\begin{aligned} t &= \phi(\bar{\boldsymbol{x}},\ \bar{t},\ \varepsilon'') = \phi[\psi(\boldsymbol{x},\ t,\ \varepsilon'),\ \phi(\boldsymbol{x},\ t,\ \varepsilon'),\ \varepsilon''] \\ &= \phi(\boldsymbol{x},\ t+\varepsilon',\ \varepsilon'') = (t+\varepsilon') + \varepsilon'' = \phi(\boldsymbol{x},\ t,\ \varepsilon' + \varepsilon''); \end{aligned}$$

and, trivially for $\boldsymbol{x}$

$$\bar{\bar{\boldsymbol{x}}} = \psi(\bar{\boldsymbol{x}},\ \bar{t},\ \varepsilon'') = \bar{\boldsymbol{x}} = \psi[\boldsymbol{x},\ t+\varepsilon',\ \varepsilon''] = \boldsymbol{x} = \psi(\boldsymbol{x},\ t,\ \varepsilon' + \varepsilon'').$$

That is, for two actions of the group defined by admissible parameter values ε' and ε'', the equivalent single action is given by the parameter value $\varepsilon' + \varepsilon''$.

2. The identity value of the parameter is $\varepsilon_0 = 0$.
3. The inverse value for any admissible value ε is $\gamma = -\varepsilon$.

4. The associativity of the group action follows in this case from the associativity of real number addition, $(\varepsilon^1 + \varepsilon^2) + \varepsilon^3 = \varepsilon^1 + (\varepsilon^2 + \varepsilon^3)$.

Further, the infinitesimal transformations are given by:

$$\xi_i(x, t) = \frac{\partial \psi_i(x, t, \varepsilon_0)}{\partial \varepsilon} = \frac{\partial x_i}{\partial \varepsilon}\Big|_{\varepsilon_0} \equiv 0; \quad i = 1, 2, \ldots, n;$$

$$\tau(x, t) = \frac{\partial \phi(x, t, \varepsilon_0)}{\partial \varepsilon} = \frac{\partial (t + \varepsilon)}{\partial \varepsilon}\Big|_{\varepsilon_0} = 1;$$

and the group operator is given by:

$$X = \tau \frac{\partial}{\partial t} - \sum_{i=1}^{n} \dot{x}_i \frac{d\tau}{dt} \frac{\partial}{\partial \dot{x}_i}.$$

2.2. Constant Discount Rate

Again we suppose that there is a single parameter $\varepsilon \in \mathbb{R}$, a simple shift in the transformation of time, and no effect on $x: \bar{t} = \phi(x, t, \varepsilon) = \varepsilon + t$ and $\psi(x, t, \varepsilon) \equiv x$. With a constant rate of discounting, the firm discounts future profits by a factor of $e^{-\rho t}$, where ρ is a constant. This leads to the conservation law,

$$\Pi + p_i \dot{x}_i = \rho \int_t^{\infty} e^{-\rho s} G(x, \dot{x}) ds,$$

indicating that the current value of profit and changes in the value of the firm must be equal to the discounted value of the firm.

2.3. Variable Discount Rate

Again we suppose that there is a single parameter $\varepsilon \in \mathbb{R}$, a simple shift in the transformation of time, and no effect on $x: \bar{t} = \phi(x, t, \varepsilon) = \varepsilon + t$ and $\psi(x, t, \varepsilon) \equiv x$. If discounting is done with a variable rate given only as $e^{-\rho(t)}$, then the conservation law is:

$$\frac{d}{dt}[e^{\rho(t)}(G + p_i \dot{x}_i)] = \frac{d\rho(t)}{dt} p_i \dot{x}_i; \; i = 1, 2, \ldots, n.$$

Sato points out that when $\rho(t) = \rho t$, so that $d\rho(t)/dt = \rho$ is a constant, we obtain the conservation law originally derived by Samuelson (1970).

2.4. Variable Discount Rate and Variable Time Transformation

Again we suppose that the group action has no effect on x, $\psi(x, t, \varepsilon) \equiv x$. Now the effect on time is given by $\tau(x, t) = 1/\rho'(t)$. Since it may be difficult to integrate $\tau(x, t)$ back to $\phi(x, t, \varepsilon)$ (not to mention integrating $\xi(x, t)$ back to $\psi(x, t, \varepsilon)$), this example illustrates one of the flexibilities of the Lie group approach to economic analysis: one can specify the action of the group through the transformations themselves—$\boldsymbol{\psi}(\boldsymbol{x},\ t,\ \boldsymbol{\varepsilon})$—or through the infinitesimal transformations—$\boldsymbol{\xi}(\boldsymbol{x},\ t)$ and $\tau(\boldsymbol{x},\ t)$.[3] In this case, Sato obtains a messy, but interpretable conservation law,

$$\frac{G + p_i \dot{x}_i|_{s=t}}{\int_t^{\infty} e^{-\rho(s)} + \rho(t)\,(G + (G + p_i\dot{x}_i)(\rho''(s)/[\rho'(s)]^2)\,)ds} = \rho'(t).$$

i.e., the ratio of the current creation of corporate income and the adjusted value of the firm equals the rate of change in the discount factor at every time t. In the special case that $\rho(t) = \rho t$, $\tau(\boldsymbol{x},\ t) = 1/\rho$ and the conservation law collapses to an earlier special case.

2.5. Variable Time Transformation and Quantity or Price Transformations

Here we pull out all the stops! Both time and the economic variables are transformed through the action of a group defined through its infinitesimal transformations:

$$\boldsymbol{\xi}(\boldsymbol{x},\ t) = \boldsymbol{x};\ \tau(\boldsymbol{x},\ t) = 1.$$

We see again the transformation of time by a simple shift, determined by the value of a single parameter, $\varepsilon : \phi(\boldsymbol{x},\ t,\ \varepsilon) = t + \varepsilon$. As to the vector of economic variables, $\boldsymbol{x}$, we note now that more than one transformation group can give rise to a particular infinitesimal transformation. Consider first the following Lie group of transformations:

$$\bar{x}_i = \psi_i(\boldsymbol{x},\ t,\ \varepsilon_1,\ \ldots,\ \varepsilon_n,\ \varepsilon_{n+1}) = e^{\varepsilon_1} x_i;\ i = 1,\ 2,\ \ldots,\ n; \tag{7}$$

$$\bar{t} = \phi(\boldsymbol{x},\ t,\ \varepsilon_1,\ \ldots,\ \varepsilon_n,\ \varepsilon_{n+1}) = t + \varepsilon_{n+1}. \tag{8}$$

Note that we have $r = n + 1$ parameters for this group (i.e., $\boldsymbol{\varepsilon} \in \mathbb{R}^{n+1}$).[4] Observe that the group properties are satisfied.

1. The composition property is illustrated by:

$$\begin{aligned}
\bar{\bar{x}}_i &= \psi_i(\bar{\boldsymbol{x}},\ \bar{t},\ \boldsymbol{\varepsilon}'') = \psi_i[\boldsymbol{\psi}(\boldsymbol{x},\ t,\ \boldsymbol{\varepsilon}'),\ \phi(\boldsymbol{x},\ t,\ \boldsymbol{\varepsilon}'),\ \boldsymbol{\varepsilon}''] = \psi_i[\bar{\boldsymbol{x}},\ t + \varepsilon'_{n+1},\ \boldsymbol{\varepsilon}''] \\
&= e^{\varepsilon'_i + \varepsilon''_i} x_i = \psi_i(\boldsymbol{x},\ t,\ \boldsymbol{\varepsilon}' + \boldsymbol{\varepsilon}'');\quad i = 1,\ 2,\ \ldots,\ n; \\
\bar{\bar{t}} &= \phi(\bar{\boldsymbol{x}},\ \bar{t},\ \boldsymbol{\varepsilon}'') = \phi[\boldsymbol{\psi}(\boldsymbol{x},\ t,\ \boldsymbol{\varepsilon}'),\ \phi(\boldsymbol{x},\ t,\ \boldsymbol{\varepsilon}'),\ \boldsymbol{\varepsilon}''] = \phi i[\bar{\boldsymbol{x}},\ t + \varepsilon'_{n+1},\ \boldsymbol{\varepsilon}''] \\
&= (t + \varepsilon'_{n+1}) + \varepsilon''_{n+1} = \phi(\boldsymbol{x},\ t,\ \boldsymbol{\varepsilon}' + \boldsymbol{\varepsilon}'').
\end{aligned}$$

2. The identity value of the parameter vector is $\boldsymbol{\varepsilon}_0 = \mathbf{0} \in \mathbb{R}^{n+1}$, i.e., the $(n+1)$-dimensional null vector.
3. The inverse value for any admissible value $\boldsymbol{\varepsilon}$ is $\boldsymbol{\gamma} = -\boldsymbol{\varepsilon} \in \mathbb{R}^{n+1}$.

[3] Of course, specifying the operator X found in Eq. (6) is equivalent to specifying the infinitesimal transformations.

[4] Note what Noether's theorem indicates for this example: If the variational problem (1) is invariant under the $(n+1)$-parameter group, then there exist $n+1$ conservation laws!

4. The associativity of the group action follows in this case from the associativity of real number addition, $(\boldsymbol{\varepsilon}^1 + \boldsymbol{\varepsilon}^2) + \boldsymbol{\varepsilon}^3 = \boldsymbol{\varepsilon}^1 + (\boldsymbol{\varepsilon}^2 + \boldsymbol{\varepsilon}^3)$.

The infinitesimal transformations for the group in Eqs. (7) and (8) are given by:

$$\xi_i(\boldsymbol{x}, t) = e^{\varepsilon_i} x_i|_{\varepsilon_0} = x_i; \quad i = 1, 2, \ldots, n; \tag{9}$$

$$\tau(\boldsymbol{x}, t) = 1. \tag{10}$$

Now consider an alternative Lie group of transformations:

$$\bar{\bar{x}}_i = \psi_i(\boldsymbol{x}, \boldsymbol{t}, \varepsilon_1, \ldots, \varepsilon_n, \varepsilon_{n+1}) = \varepsilon_i x_i; \quad i = 1, 2, \ldots, n; \tag{11}$$

$$\bar{t} = \phi(\boldsymbol{x}, \boldsymbol{t}, \varepsilon_1, \ldots, \varepsilon_n, \varepsilon_{n+1}) = t + \varepsilon_{n+1}. \tag{12}$$

Again we have $r = n + 1$ parameters for the group, and the transformation of time is once again the simple one-parameter shift. Observe again that the group properties are satisfied.

1. The composition property is illustrated by:

$$\begin{aligned}
\bar{\bar{x}}_i &= \psi_i(\bar{\boldsymbol{x}}, \bar{\boldsymbol{t}}, \boldsymbol{\varepsilon}'') = \psi_i[\boldsymbol{\psi}(\bar{x}, t, \boldsymbol{\varepsilon}'), \phi(\boldsymbol{x}, \boldsymbol{t}, \boldsymbol{\varepsilon}'), \boldsymbol{\varepsilon}''] = \psi_i[\bar{\boldsymbol{x}}, t + \varepsilon'_{n+1}, \boldsymbol{\varepsilon}''] \\
&= \varepsilon'_i \, \varepsilon''_i x_i = \psi_i(\boldsymbol{x}, \boldsymbol{t}, \boldsymbol{\varepsilon}'''); \, i = 1, 2, \ldots, n; \\
\bar{t} &= \phi(\bar{x}, \bar{\boldsymbol{t}}, \boldsymbol{\varepsilon}'') = \phi[\boldsymbol{\psi}(x, t, \boldsymbol{\varepsilon}'), \phi(\boldsymbol{x}, \boldsymbol{t}, \boldsymbol{\varepsilon}'), \boldsymbol{\varepsilon}''] = \phi(\bar{x}, t + \varepsilon'_{n+1}, \boldsymbol{\varepsilon}'') \\
&= (t + \varepsilon'_{n+1}) + \varepsilon''_{n+1} = \phi(\boldsymbol{x}, \boldsymbol{t}, \boldsymbol{\varepsilon}'')
\end{aligned}$$

where $\varepsilon'''_i = \varepsilon'_i \varepsilon''_i, \quad i = 1, 2, \ldots, n$; and $\varepsilon'''_{n+1} = \varepsilon'_{n+1} + \varepsilon''_{n+1}$.

2. The identity value of the parameter vector is

$$\boldsymbol{\varepsilon}_0 = [1 \; 1 \ldots 1 \; 0]^{\mathrm{T}} \in \mathbb{R}^{n+1}$$

3. The inverse value for any admissible value $\boldsymbol{\varepsilon}$ with $\varepsilon_i \neq 0$ for $i = 1, 2, \ldots, n$, is

$$\gamma = \left[\frac{1}{\varepsilon_1} \; \frac{1}{\varepsilon_2} \cdots \frac{1}{\varepsilon_n} - \varepsilon_{n+1}\right]^{\mathrm{T}}$$

4. The associativity of the group action follows in this case from the associativity of real number multiplication and addition.

The infinitesimal transformations for the group in Eqs. (11) and (12) are given by:

$$\xi_i(\boldsymbol{x}, t) = x_i; \quad i = 1, 2, \ldots, n; \tag{13}$$

$$\tau(\boldsymbol{x}, t) = 1. \tag{14}$$

Compare the infinitesimal transformations in Eqs. (9) and (10) with those in Eqs. (13) and (14): they are identical! Thus, different groups of transformations may have the same infinitesimal transformations (and, therefore, the same operator, X).

3. Concluding Remarks

Previous explorations into conservation laws in economics were of a macroeconomic nature. The results were derived in highly specialized circumstances. While those examples are nevertheless interesting from a mathematical perspective, the *micro* economic conservation laws introduced in Sato's paper are fascinating for a fundamental economic reason: they can be derived as the result of rational, optimizing behavior.

There are potentially as many microeconomic conservation laws as there are different types of optimizing behavior. While profit-maximization may still be the leading long-run objective sought by individual managers or firms, it is not necessarily the only behavioral model over a shorter planning horizon. For alternative objectives, one must only modify the original variational problem in Eq. (1) and the rest of Sato's paper still follows. As Sato notes in his introduction, there are many criteria or measures of corporate performance. Profits, dividends, and price-to-earnings-ratios are but some of the possible measures. Some firms may target earnings growth or market share growth over an n-year period, while maintaining a given level of total profit (as a minimum), or while maintaining a given quarterly divided. With all of the potential measures of corporate performance, there is a large number of permutations and combinations of these performance criteria, and each permutation or combination could potentially reveal a microeconomic conservation law. We have only to assemble the data and search. As Paul Samuelson put it in the foreward of Sato's monograph on the application of Lie groups of transformations (Sato, 1981, 1999), "the ball is now in our court."

References

Mitchell, T., Primont, D., 1991. Functional forms for technical change functions. Journal of Productivity Analysis 2, 143–152.

Noether, E., Invariant variationsprobleme. Nachr. Akad. Wiss. Göttingen, Math.- Phys. Kl. II, 1918, pp. 235–257.

Ramsey, F., 1928. A mathematical theory of saving. Economic Journal 38, 543–559.

Samuelson, P.A., 1970. Law of conservation of the capital-output ratio. In: Proceedings of the National Academy of Sciences. Applied Mathematical Science 67, 1477–1479.

Sato, R., Theory of Technical Change and Economic Invariance: Application of Lie Groups. New York: Academic Press, 1981. Reprinted, Cheltenham, Edward Elgar, UK, 1999.

Chapter 10

EMPIRICAL TESTS OF THE TOTAL VALUE CONSERVATION LAW OF THE FIRM

1. Introduction

At the macro level, the efficient operation and performance of nations are analyzed on the assumption that they are carried out in ways that maximize per capita consumption over the long-term. The income to national wealth ratio is known as a theoretical tool derived from an economic conservation law (Sato (1999, 2002)).

It is also shown that at the microeconomic level, economic conservation laws could apply to evaluate the corporate behavior. From the perspective of pure theoretical economics, a corporation may be analyzed on the assumption that it is operating to maximize profits over a certain period under the given conditions.

In reality, however, production costs and the exact workings of R&D and other factors are uncertain. Since the numerical values of certain crucial parameters are unobservable, it is impossible to judge whether a corporation's reported profits are being maximized or not. Under such circumstances it is not whether the company has achieved maximum profits but whether its profits has shown positive growth that managers regard as the primary criterion of success. In short, there is a tendency to regard someone as a "good" manager as long as profits are growing even if he/she fails to achieve maximum profits under the given conditions.

This paper applies a conservation law to individual industries and companies, and empirical tests are presented to show the possibility that there are ways to evaluate manager behavior relative to the profit-maximization goal. Although the results are tentative because of the constraints of data availability, they represent the first extensive investigation to apply a conservation law using Japanese corporate data.

In conducting the empirical analysis, a constant discount rate is assumed. As a first approximation, this appears to be reasonable. Such a result is particularly interesting because the data include the period of the "asset-price bubbles". Further, a systematic relationship between the stability of the ratio of income creation to the value of the firm, which should be equated to the discount rate applicable to a firm, and a relevant measure of corporate performance, such as survey rankings, has been suggested for certain classes of firms.

The paper begins with the theoretical derivation of the conservation laws that are applied to corporate long-term profit maximization behavior. With this context, empirical evidence using the Japanese corporate data is examined. The relationship between traditional valuation theory of the corporate and valuation based on the optimizing behavior is then presented. The final section discusses the implications of the analysis and an agenda for future research.

2. Conservation Law and Firm Valuation in Theory

To consider a firm's profit maximization, the following profit function is formulated:

$$J(x) = \int_a^b \Pi(t,\ x(t), \dot{x}(t))dt \tag{1}$$

where $x = (x^1(t), \ldots, x^n(t))$, which equals the vectors of functions of quantities and prices, and $\dot{x}; (t) = dx/dt$.

For $J(x)$ to be maximized, the necessary condition for the optimal solution is that the Euler-Lagrange equations vanish, thus;

$$E_i = \frac{\partial \Pi}{\partial x^i} - \frac{d}{dt}\left(\frac{\partial \Pi}{\partial \dot{x}^i}\right) = 0 \tag{2}$$

Using a system of Lie group transformations and Noether's Invariance Principle, we can derive the following general form of Conservation Laws for evaluating corporate performance.

$$\left(\Pi - \sum_{i=1}^{n} \dot{x}^i \frac{\partial \Pi}{\partial \dot{x}^i}\right)\tau(x,\ t) + \sum_{i=1}^{n} \frac{\partial \Pi}{\partial \dot{x}^i}\xi^i(x,\ t) = \Phi(x,t) \tag{3}$$

where $\tau(x,t)$ and $\xi(x,t)$ are infinitesimal transformations associated with time t and x respectively, and $\Phi(x,t)$ is a null term.[1]

If we define $p^i = -\partial\Pi/\partial\dot{x}^i$ = supply price of x^i, the above implies that (profit $+\sum$ (additional value created)) $\times$ time transformation $+\sum$ (supply price of x^i) $\times$ x^i transformation = null term. This is too general to be useful for testing corporate performance.

There are several cases where τ and ξ take special values and/or the profit function has a special form.

[1] See Sato (2004), equation (11) or (11′).

Suppose that there is no x^i transformation, namely $\xi^i = 0$ and that the time transformation is constant, thus, $\tau = \text{const.} = 1$. If the profit function has a familiar form such as $\Pi = e^{-\rho t}G(x(t),\ \dot{x}(t))$ where ρ is discount rate, then equation (3) can be expressed as

$$\left(G(x(t),\ \dot{x}(t)) + \sum \pi^i \dot{x}^i\right) = \Phi = \rho \int_t^\infty e^{-\rho(s-t)} G(x,\ \dot{x})ds \tag{4}$$

where $\pi^i = e^{\rho t}p^i$ and Φ is derived from $-\int_t^\infty (\partial \Pi / \partial s)\, ds$[2].

By transforming equation (4) it may be more convenient to write it as,

$$\rho = \frac{G + \sum \pi^i \dot{x}^i}{\int_t^\infty e^{-\rho(s-t)} G(x, \dot{x})ds}. \tag{5}$$

The equation (5) says that the current value of profit, G, and changes in the value of the firm, $\sum \pi^i \dot{x}^i$, divided by the discounted value of the firm, $\int_t^\infty e^{-\rho(s-t)} G(x,\ \dot{x})ds$, must always be equal to the discount rate ρ. This is the conservation law that holds for all t, if under the assumption of a constant discount rate, the firm is seeking to maximize long-run profits.

To the extent that the concept of conservation laws is useful in evaluating the actual performance of a firm, we should be able to construct a model for empirical analysis. The problem is that it is not possible to know the exact value of the discount rate of a particular firm. Assuming that the firms are working under the condition of a constant discount rate, we can explore the implication of equation (5). In a preliminary investigation using this approach on US data, Sato (2004) found that the ratio of income creation to the value of the company was more stable for the aircraft industry than the automobile industry (Mitchell, 2004).

If markets value firms correctly, we can test whether conservation laws hold for firms by using data from their financial statements and market prices of their securities and shares. Because the theory predicts that if a firm optimizes profits, the derived ratio of income creation to the value of the firm should be relatively stable, we should observe that the standard deviation of the calculated ratio is small for a firm making optimizing decisions, and thus it should be negatively correlated with some relevant measure of the performance of a firm, such as analyst evaluations or ranking data released by an independent survey institution. Based on this idea, we conducted an empirical analysis, the results of which follow.

3. How Do the Japanese Data fit the Model?

3.1. Calculating the Ratio

If a firm engages in long-run profit maximizing operations in whatever form it may be, we may be able to observe the outcome of this behavior from the published corporate financial data and market prices of their securities and shares. Because we do not know the exact value of the discount rate that each

[2] See Sato (2004), equation (11′).

firm regards as its given rate, we must infer the value of discount rate according to equation (5).

In this section, a series of the ratio is derived using the following equation (6) and then these are subjected to a regression analysis. Suppose that a corporation issues only equity, then we can interpret the variables in equation (5) as: G is the current value of profit, $\int_t^{\infty} e^{-\rho(s-t)} G(x, \dot{x})ds$ is the current value of a firm, which equals the market capitalization of the firm; and $\sum \pi^i \dot{x}^i$ is the change in the current value of the firm measured as the change in market capitalization. In reality, most firms have debt and such debt are to be included in the calculation.

We first construct the series of the value of the ratio from the financial statement data and share prices, based on the above interpretation and test the prediction of corporate behavior.

To do this, we define the variables for firm i during or at the end of period t as follows:

R_{it}: current profit, V_{it}: total value of the firm at market prices, equal to $E_{it} + B^*_{it}$, E_{it}: market capitalization of equity, B^*_{it}: market value of debt. Although it is better if debt is evaluated at market prices, it is not easy to obtain appropriate values, so thier book, B_{it}, is used in our analysis. This is not a problem for most well-managed firms, as their credit risk premium is presumably quite small.[3]

To avoid annual idiosyncratic variations in profits and firm values, a three-year moving average is used to obtain the series of the ratio which we call the calculated discount rate $\hat{\rho}_{it}$. That is,

$$\hat{\rho}_{it} = \frac{1}{3} \sum_{k=t-2}^{t} \left(\frac{R_{ik} + \Delta V_{ik}}{V_{ik}} \right) \tag{6}$$

where $\Delta V_{ik} = V_{i,k} - V_{i,k-1}$.[4] *We define* $\sigma_{\hat{\rho}_{it}}$ as the sample standard deviation of $\hat{\rho}_{it}$. We use a three-year moving average when constructing a data series in order to capture the trends under which firms are operating. The length chosen is somewhat arbitrary, and if longer underlying data were available, other appropriate smoothing techniques might be used.

3.2. The Data

The source of our sample firms is firms listed on the first section of Tokyo Stock Exchange. Their financial statements are published regularly.[5] These financial statements are collected and compiled by Toyokeizai (a Japanese publisher) as Corporation Financial Statements in an electronic database format (Toyokeizai, 2003).

[3] We include interest-bearing trade credits, as such credits can be significant in Japan, and the data are disclosed in financial statements. This means B_{it} here is broader than what most US analysts call "debt".

[4] Note that the ratio being calculated is not the same as the return on assets or similar financial-analysis measures.

[5] Annual reports of public companies are published by the Printing Office of the Government. Database compiled by Toyokeizai is a historical collection of these published annual reports.

Table 1. Calculated Values of the Ratio $\hat{\rho}$: Manufacturing Industries.

Industrial category	Number of firms	Mean of $\hat{\rho}$	Standard deviation of $\hat{\rho}$
Construction	124	0.0595	0.1044
Food Processing	63	0.0588	0.1060
Chemical	110	0.0522	0.1056
Iron and Steel	28	0.0359	0.1247
Machinery	95	0.0494	0.1234
Electric Appliances	140	0.0666	0.1402
Automobile and Related	65	0.0567	0.0987
Other Manufacturing	29	0.0783	0.1368
Manufacturing Total	880	0.0566	0.1166

Note:

1. Numbers are industry averages of mean and standard deviation of $\hat{\rho}_{it}$ for firms belonging to each industrial category;
2. Because some companies changed their accounting period, we allow one missing observation which results in 19 observations during the period between March 1981 and March 2003. Such missing observation occurs when a corporation changes their accounting period, for example from starting in June to starting in April.

Firms with more than 19 annual data points in the 22-year period, March 1981 through March 2003, are included in our data set.[6] If a mean value of calculated $\hat{\rho}_i$ is negative, the firm is excluded because a negative discount rate is hard to reconcile with theory. In practice, current profits can have negative values for several years, and share prices can decline for prolonged periods.

The resulting sample is 1,170. For the regression analysis, only some of firms are used, with the exact sample size varies depending on the model and explanatory variables used.

First, we look at the calculated series of mean and standard deviation of $\hat{\rho}_i$. These are presented in Tables 1 and 2 for 14 major industries as illustrations of the 29 industries in the dataset.

The average values of the mean of the calculated ratio, $\hat{\rho}_i$, are between 5% and 8% for all but three industries, and only services (9%) and iron and steel (3.6%) are truly outliers. This seems consistent with general macroeconomic variables. The sample standard deviations are 10% to 14%, with only gas and electric (7.8%) the real outlier.

Annual variations of $\hat{\rho}_i$ were limited in the range of one standard deviation from their mean for most of the samples. Thus, the idea of constant ρ may not be a bad approximation. To explore the relationships more systematically, we proceed to the regression analysis.

3.3. The Regression

Estimation is conducted by ordinary least squares method, using the following equation.

[6] So that firms that change their accounting period can be included, the data points can be over a 20-year period. That is, one discontinuity is allowed.

TABLE 2. Calculated Values of the Ratio $\hat{\rho}$: Non-Manufacturing Industries.

Industrial category	Number of firms	mean of $\hat{\rho}_i$	standard deviation of $\hat{\rho}_i$
Gas & Electricity	17	0.0707	0.0778
Transportation	31	0.0610	0.1014
Wholesale	89	0.0565	0.1033
Retail	47	0.0711	0.1224
Real Estate	22	0.0733	0.1381
Services	39	0.0914	0.1378
Non-Manufacturing Total	290	0.0678	0.1143

Note:

1. Numbers are industry averages of mean and standard deviation of $\hat{\rho}_{it}$ for firms belonging to each industrial category;
2. Because some companies changed their accounting period, we allow one missing data which results in 19 observations during the period between March 1981 and March 2003;
3. No telecommunication company has enough long history of financial data, thus we miss that industry from our data set.

$$\sigma_{\hat{\rho}_i} = \beta_1 PI_i + \beta_2 \left(\frac{L}{V}\right)_i + \beta_3 DI_i + \varepsilon_i \tag{7}$$

where PI_i is any performance indicator of a firm i, ε_i is an error term; and DI denotes the industry dummy variable, which takes the value one if firm i belongs to industry I and zero otherwise. The dummy variables of $D1$ through $D28$ cover 29 industry types, of which 10 industries belong to the non-manufacturing sector. In this paper, only non-financial corporations are considered.

$(L/V)_i$ is introduced to adjust for possible effects from the level of leverage. This equals the average ratio of total liabilities bearing interest charges to the value of the firm. If the discount rate is constant, then when optimizing firms are maintaining a constant value of the ratio on the right hand side of equation (5), we except β_1 to be negative and, possibly, β_2 to be positive if leverage disturbs the optimization.

How do we know whether a firm is really maximizing long-term profits? That is a key question in the empirical tests, and the answer relates to the choice of performance indicator PI_i. Several possibilities are used.

3.4. Regression Results

We first tested the model with the Nikkei Performance Index as PI_i. This index is constructed annually and published, usually in September, by Nikkei, a major Japanese media company that issues daily newspapers (The Nihon Keizai Shimbun among them), and is independent of the other variables in the equation. Nikkei ranks firms according to four items, with a total maximum score of 1,000 for a firm. The items and weights for the 2003 ranking are: profitability (34.7%), size (25.4%), stability (degree of leverage, etc) (20.7%), and growth potential

TABLE 3. Regression Results (1).

	PIi	$(L/V)_i$	$\bar{R}^2$	N
All available data	−0.0889 [0.002]	−0.1424 [0.000]	0.2460	295
Manufacturing	−0.0897 [0.005]	−0.1386 [0.006]	0.1953	223
Selected firms:all categories	−0.0897 [0.003]	−0.1308 [0.000]	0.2311	272

Note:

1. N is number of observations. Numbers in brackets are p-values;
2. Industry dummy variables are included whose coefficients are not reported in the table. Except Mining, Oil and Coal, Iron and Steel, and Metals, these dummies variables are statistically significant at either 5% or 1% level;
3. "Selected firms" means the sample set that have less than 10 negative $\hat{\rho}_{it}$ for firm i in their data history among 19 or 20 total entries.

(19.2%).[7] In our regressions, the Nikkei Performance Index is re-based by dividing by 1,000, so the highest score is one. Since we use financial data until year 2002, the following 2003 ranking is used as PI_i in equation (7).

Table 3 shows the regression results. The Nikkei releases ranking only for top 600 firms, so the number of firms that have both a Nikkei ranking and a complete history of financial data is less than 300. From the table, we can confirm the negative coefficients for PI_i for all three cases at a statistical significance level of 1%. That is, in the data collected here, the higher the latest performance index, the lower the volatility of the calculated ratio in a firm's preceding history.

The data set labeled as "selected firms" are firms which have less than 10 negative values of the calculated ratio as $\hat{\rho}_{it}$. The result is basically same for this group as for the larger group.

For the leverage variable, the negative coefficients are statistically significant by $(L/V)_i$. One possible explanation why leverage matters is that the expected higher volatility of the equity part of total value varies directly with the degree of leverage, and this causes higher overall variability. If this is true, then high leverage should be associated with a high volatility of the ratio. However, the sign of the estimated coefficients contradicts this story.

Table 4 summarizes the regression results for an alternative performance index, return on assets ($(R/V)_i$). Note that this allows for much larger samples. An average value of return on total assets does not appear to be a plausible explanatory variable, as shown by statistical insignificance of the results. This is true for all samples, as well as for sub-groupings. The estimated coefficient of the leverage variable again turns out to be negative and statistically significant. "Selected firms" refers to the same data set as in Table 3. Here, the estimated coefficient of the average return on assets has a positive sign with statistical significance and the leverage variable does not seem to affect the size of volatility in the calculated ratio.

The growth of total assets over the sample period also was tested. Specifically, three-year averages of the total assets of firm i, (V_i), for the first-three and

[7] For a fuller explanation of the index, see September 27, 2003 issue of the Nihon Keizai Shimbun.

TABLE 4. Regression Results (2).

	$(R/V)_i$	GR_i	$(L/V)_i$	R^2	N
All industries (1)	0.0338 [0.644]		−0.0766 [0.000]	0.1289	1112
Manufacturing	0.0327 [0.679]		−0.0550 [0.031]	0.1068	834
Non-manufacturing	0.0971 [0.605]		−0.0954 [0.002]	0.1785	277
Selected firms	0.4683 [0.008]		−0.0705 [0.122]	0.2252	272
All available data with Nikkei index	0.3991 [0.021]		−0.0779 [0.045]	0.2349	295
All industries (2)		0.0014 [0.000]	−0.0714 [0.000]	0.1729	990

Note:

1. N is number of observations. Numbers in brackets are p-values.
2. Industry dummy variables are included whose coefficients are not reported in the table. Among them, only dummy variable for the wholesale and retail industry is statistically significant at the 5% level.
3. "Selected firms" means the sample set that have less than 10 negative $\hat{\rho}_{it}$ for firm i in their data history among 19 or 20 total entries.

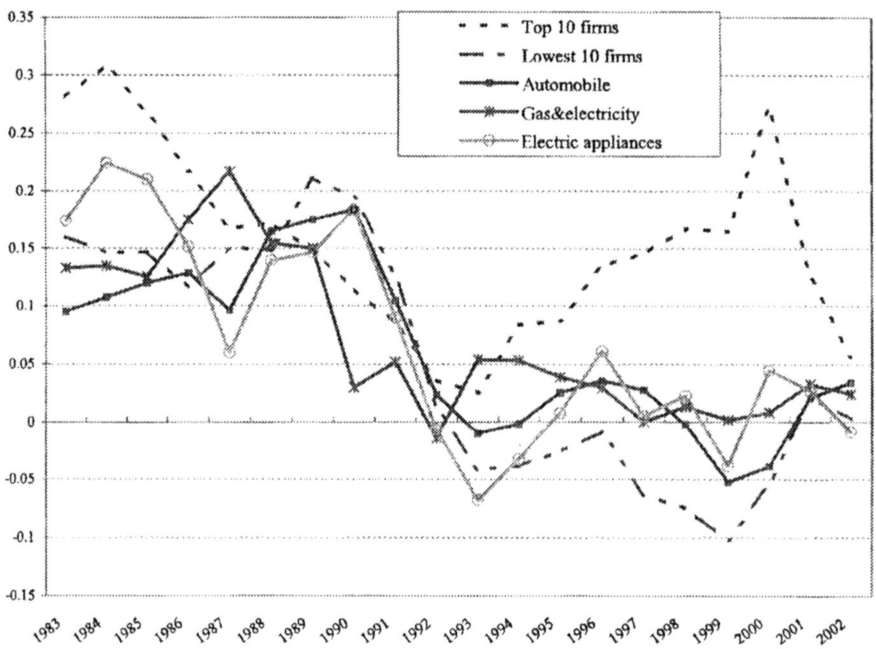

FIGURE 1. Changes of the Ratio: Top and Lower Classes of Firms and Industrial Averages.
Notes:

1. The top 10 and lowest 10 firms are selected from the data set that have both 2003 Nikkei Performance Index and enough history of financial data as shown in Tables 1 and 2. Two dashed lines correspond to these groups: higher one shows the top 10 average and lower one shows the lowest 10 average.
2. The bold lines correspond to the average values of the ratio for three industries; automobile, gas and electricity and electric appliances respectively.

last-three years of the data period -that is, 1980–1982 and 2000–2002-are calculated, and a growth ratio between the two points, GR_i, is obtained as a performance index.[8] This index has a positive sign with statistical significance, as shown in the last row of Table 4.

In figure 1, the calculated values of the ratio are shown for two classes of firms and average values for three industries. The top-10 and lowest-10 firms in the 2003 Nikkei Performance Index (NPI) both perform relatively well given that even the lowest in this data set are among the better firms in the set of all firms. However, we notice a difference in the plots of the average values of the calculated ratios of these two groups.

The calculated ratios for the industries show more stable behavior than the plots of the sub-sets of NPI companies. The data for the individual firms seem to be affected by short-term fluctuations of the factors surrounding them.

Regarding the size of the estimated discount rate, the top-performing firms appear to have a relatively higher value, which implies a greater value of the discount rate if they engage in maximization behavior. Despite the popular view that Japanese firms experienced structural changes when the bubble crashed, the top-performing firms are still growing, as reflected in higher values of their ratios. For most of the japanese firms, the fact that the plots of industry averages are lower in the 1990s than in the 1980s suggests that firms' behavior in the 1990s changed from the 1980s, and thus different values of the discount rate might apply.

3.5. Data Issues

The regression results are inconclusive. One of the problems is the choice of the performance variable.

Using the Nikkei index shows the possibility that better-managed firms are related to low volatility of the ratio, which is consistent with our assumption of a constant discount rate. Indices constructed from the financial-statement data suggest that, the greater the values of either return on assets or the growth of total assets, the higher the volatility of the ratio. In this context, it is not plausible to interpret the results as saying well-managed firms are operating with a constant discount rate. Further sophisticated index or variable to distinguish firms engaging in long-term profit maximization from others need to be explored.

Another data issue is survivorship bias. In constructing our data set, such bias is unavoidable, as we need a long history of company data. Thus, companies that are short-lived because of poor management or any reasons are excluded from the sample. Put differently, the tests in this paper are not ideal for identifying specific firms with maximizing operations. However, we are more concerned with the relevancy of the assumption of a constant discount rate.

As shown in Tables 1 and 2, comparative analysis among different industries is of great interest. However, to choose the right variable to measure industrial performance is much harder than to choose the right variable to measure

[8] In some cases-for example, because of a change in fiscal years-the data are for slightly different periods, but always cover at least 20 years.

individual corporate performance. That is, while performance indicators for individual firms are available, we have not found a performance indicators for industries that is suitable for industry-level regression.

4. Relating to Traditional Corporate Valuation Theory

In traditional corporate valuation theory, the discount rate is related to the variability of future cash flows if uncertainty prevails. The discount rate is the reward for the business risk that investors undertake, and this is reflected in the price of a firm's shares. In a certain world, the discount rate is equal to the interest rate as the reward for time.

In the optimization context under certainty, the necessary condition requires that, at any moment of time, equation (5) holds. From an empirical viewpoint, the question is how the managers could know the given value of their discount rate. In this paper, we interpret values from corporate financial statement as clues to understanding the real value of a firm's discount rate in equation (5). There may be more hints that a firm reveals its discount rate through its operating behavior.

In the previous section, we measured the volatility of calculated value of the ratio $\hat{\rho}_i$. Because of the way that $\hat{\rho}_i$ s are constructed, some may worry that this simply measures the volatility of the market value of equities. To check this concern, we calculated the correlation coefficients between the volatility of market capitalization of firm i and that of $\hat{\rho}_i$. It is -0.057, which means the volatility of $\hat{\rho}_i$ in this paper captures an aspect of corporate behavior different from uncertain cash flows.

The issue is how to abstract the real world and construct a convincing model as a good approximation of firm behavior. Traditional valuation theory under uncertainty provides a very good basis to explain the variation in return for individual firms. In this context, firms are given their discount rate from capital markets as an opportunity cost of their investment. The mechanism to force the managers to optimize is the forces of capital markets. Thus, the surviving firms are well assumed to be optimizing.

On the other hand, given the relevant value of the discount rate, it is important to examine whether a firm has been optimizing its operations even under the assumption of certainty. From the viewpoint of this analytical purpose, we are interested in longer-term trends in corporate behavior, rather than daily fluctuations in the market valuation of the firm. In this regard, we have shown a possibility that the assumption of a constant discount rate is not necessarily contradicted by the data.

5. Concluding Remarks

We have shown evidence that the assumption of a constant discount rate and profit-maximizing behavior of a firm is consistent with observed data for many of the top-performing firms in Japan during the period of 1980 through 2002. Although the results are tentative because of the limited data period and data availability, they suggest that it is promising to extend the notion of conservation laws to the industry or firm level.

The valuation of a firm is far from a simple and easy exercise. A measure related to a conservation law could be used together with traditional standards of measurement found in management, accounting and finance theory. Indeed, the multiplicative valuations may serve to further the understanding of corporate behavior.

From the viewpoint of the integration of optimization theory and uncertainty surrounding corporate decision making, the model under uncertainty is of interest. Toward this end, the "variable" discount rate case could be examined in more detail and developed to allow empirical investigations.

References

[1] Mitchell, T., 2004, Conservation laws for microeconomists! Comments on "Economic Conservation Laws as Indices of Corporate Performance" by Ryuzo Sato. Japan and the World Economy 16, 269–276.

[2] Sato, R., 1999, Theory of technical change and economic invariance: Application of Lie groups. New York: Academic Press, 1981. Reprinted, Cheltenham, UK: Edward Elgar.

[3] Sato, R., 2002, Optimal Economic Growth: Test of Income-Wealth Conservation Laws in OECD Countries, Macroeconomic Dynamics 6, 548–572.

[4] Sato, R., 2004, Economic conservation law as indices of corporate performance, Japan and the World Economy 16, 247–267.

[5] Toyokeizai, 2003, Toyo Keizai Data Disc-Kaisha Zaimu Karute.

Chapter 11

HARTWICK'S RULE AND ECONOMIC CONSERVATION LAWS

1. Introduction

In a series of influential papers starting with Solow (1974), Hartwick (1977) and Dixit et al. (1980), the investment rules for intergenerational equity have been proposed. Solow (1974) applies the max-min principle to the intergenerational problem of optimal capital accumulation, and suggests that the max–min principle requires consumption per capita to be constant over time.

Hartwick (1977) investigates the specific investment rule for a constant utility path; society invests all rents from exhaustible resources in reproducible capital goods—known as Hartwick's rule. This rule keeps the total value of net investment equal to zero and is shown to be sufficient for a constant utility path. Dixit et al. (1980) generalize Hartwick's rule and suggest that keeping the present discounted value of total net investment constant over time is necessary and sufficient for constant utility.

Economic conservation laws provide constancy and invariance along an optimal path. The presence of constancy motivates investigation of the relationship between conservation laws and Hartwick's rule. In this connection, we examine whether Hartwick's rule can be derived from one of the conservation laws, and furthermore whether it can be extended to obtain a more general rule. It will be shown that, in fact, Hartwick's rule is a special case when the Hamiltonian of the optimal model (the simplest form of the conservation law) is identically equal to

The authors wish to acknowledge helpful comments on earlier version by Paul A. Samuelson and A. Dixit.

zero. In general, the Hamiltonian need not be equal to zero and this fact will enable us to derive another policy rule for investment.

2. A Simple Model for Capital Accumulation with Exhaustible Resources

We start with a simple model for optimal capital accumulation with extraction costs of resources. We acknowledge a possibility that extraction of resources is costly. The extraction cost is assumed to be a proportionate to the extracted amount of exhaustible resources. Then the aggregate income is spent by consumption expenditure, investment spending on physical capital, and extraction cost of natural resources,

$$Y(t) = C(t) + \dot{K}(t) + aR(t) \tag{1}$$

where $Y(t)$ is the output at time t, $C(t)$ the consumption, $\dot{K}(t)$ the time derivative of reproducible capital $K(t)$ or investment, $R(t)$ the extracted amount of exhaustible resources, and $a > 0$ the extraction cost. Reproducible capital and natural resources are required in producing output.

We assume the constant-return-to-scale production function,

$$Y(t) = f(K(t),\ R(t)). \tag{2}$$

In this situation, the society's problem is to find the largest constant consumption with a feasible pattern of resource use and investment over the infinite time horizon. According to Solow and Wan (1975), the problem can be restated in an equivalent form. It is to choose a constant C, and find the feasible pattern of resource use and investment that minimizes the cumulative use of resources over infinite time. Then adjust the size of C until the optimized resource use is exactly equal to the stock of resources. Thus, the stated problem is to minimize

$$\int_0^\infty R(t)dt \tag{3}$$

subject to

$$\dot{K}(t) = f(K(t),\ R(t)) - aR(t) - C \tag{4}$$

and

$$\dot{X}(t) = -R(t). \tag{5}$$

Eq. (5) can be presented by an iso-perimetric constraint

$$X(t) = S - \int_0^t R(s)ds,$$

where S is the total stock of exhaustible resources known at the beginning of time. By differentiating the above equation with respect to time t, we can obtain Eq. (5).

Then Lagrange function can be written as

$$\mathscr{L} = R(t) + q_1[\dot{K}(t) - f(K(t),\ R(t)) + C + aR(t)] + q_2[\dot{X}(t) + R(t)]. \tag{6}$$

Solving for minimum, we obtain the following Euler–Lagrange first-order conditions,

$$\mathscr{L}_K - \frac{d}{dt}\mathscr{L}_{\dot{K}} = 0,\ q_1 f_K + \dot{q}_1 = 0, \tag{7}$$

$$\mathscr{L}_R = 0,\ q_1(f_R - a) - q_2 = 1, \tag{8}$$

$$\frac{d}{dt}\mathscr{L}_{\dot{X}} = 0,\ \dot{q}_2 = 0. \tag{9}$$

By using Eqs. (7)–(9) to eliminate q_1 and q_2, we obtain Hotelling's rule,

$$f_K = \frac{(d/dt)(f_R - a)}{f_R - a} \tag{10}$$

This means that along an optimal path the proportional rate of change of the rent from using the natural resources should always equal the level of the marginal productivity of reproducible capital.

3. Hartwick's Rule for Constant Consumption

In his paper, Hartwick (1977) suggests that a rule, keeping investment equal to the rents from exhaustible resources, yields a path of constant consumption.

$$\dot{K}(t) = (f_R - a)R(t). \tag{11}$$

Differentiating Eq. (11) with respect to time t and using Hotelling's rule, we obtain

$$\ddot{K}(t) = f_K(f_R - a)R(t) + (f_R - a)\dot{R}(t). \tag{12}$$

Since $(f_R - a)R(t) = \dot{K}(t)$ of Eq. (11), Eq. (12) can be rewritten as

$$\ddot{K}(t) = f_K\dot{K}(t) + (f_R - a)\dot{R}(t). \tag{13}$$

From the budget constraint, Eq. (4), the rate of change in consumption can be represented by

$$\dot{C}(t) = f_K\dot{K}(t) + f_R\dot{R}(t) - \ddot{K}(t) - a\dot{R}(t). \tag{14}$$

With Hartwick's rule which gives rise to Eq. (13), we can easily verify that $\dot{C}(t) = 0$. It means that Hartwick's rule is a sufficient condition for a constant consumption path. Now we ask ourselves if Hartwick's rule is the only rule for a constant consumption path, or if there exist other policy rules. In the next section, we use conservation laws to determine if there are other policy rules for a constant consumption path.

4. A Conservation Law for Constant Consumption

In this section, we examine whether Hartwick's rule can be derived from one of the conservation laws and whether there is another policy rule or rules for a constant consumption. Also, we investigate the relationship between Hartwick's rule and conservation laws. We use the simple model presented in Section 2 to derive a conservation law. When the variational problem in continuous time can be written as the model with Eqs. (3)–(5), since the integral does not contain t, the Hamiltonian is constant.

$$H = \mathscr{L} - \dot{K}\mathscr{L}_{\dot{K}} - \dot{X}\mathscr{L}_{\dot{X}} = -A(= \text{constant}). \tag{15}$$

Applying this to our specific model, we can obtain a simple conservation law as follows:

$$q_1(\dot{K} - (f_R - a)R) = A \tag{16}$$

or

$$q_1(C + Rf_R - f) = -A,$$

which is equivalent to

$$q_1(C - Kf_K) = A \tag{17}$$

if $f(K, R)$ is first degree homogeneous with respect to K and R.

Eqs. (16) and (17) can be written, respectively, as

$$\dot{K} = (f_R - a)\left(R + \frac{A}{1 + q_2}\right), \tag{18}$$

$$f_K K = C + (f_R - a)\frac{A}{1 + q_2}, \tag{19}$$

where A is constant. Also, q_2 is constant in Eq. (9) and $q_1 = (1 + q_2)/(f_R - a)$ from Eq. (8). Because A is any constant, when A is equal to zero, we have a special policy rule which is the same as Hartwick's rule. That is, when A is equal to zero in Eq. (6), since q_1 should not be equal to zero, the investment should equal to the rent from the extraction of resources,

$$\dot{K} = (f_R - a)R.$$

The simple conservation law, Eq. (16), is general, in the sense that the conservation law includes Hartwick's rule as a special policy rule.

Now we check whether the conservation law is a sufficient condition for a constant consumption path. Differencing Eq. (18) with respect to time t and using Hotelling's rule, we obtain

$$\ddot{K} = f_K(f_R - a)R + (f_R - a)\dot{R} + \frac{A}{1+q_2} f_K(f_R - a). \tag{20}$$

Using Eq. (18), Eq. (20) becomes

$$\ddot{K} = f_K\dot{K} + (f_R - a)\dot{R}. \tag{21}$$

From Eq. (4), $\dot{C}$ can be written as

$$\dot{C} = f_K\dot{K} + (f_R - a)\dot{R} - \ddot{K}. \tag{22}$$

By substituting Eq. (21) into Eq. (22), we can verify that

$$\dot{C} = 0.$$

Therefore, the conservation law is a sufficient condition for a constant consumption and is more general policy rule than Hartwick's rule.

5. The Elasticity of Substitution and the Hotelling Differential Equation

It may be of some interest to relate Hotelling's rule to the structure of production technology. More specifically, we are interested in how Hotelling's rule is influenced by the elasticity of factor substitution. With the assumption of the constant-return-to-scale production function, the elasticity of factor substitution, $\sigma(k)$ is expressed as

$$\sigma(k) = \frac{g'(k)[g(k) - kg'(k)]}{-kg(k)g''(k)}, \tag{23}$$

where $g(k) - f(K/R,\ 1)$, $k = K/R$, $g'(k) = f_K$ and $g(k) - kg'(k) = f_R$. Then, Hotelling's rule can be written as

$$g'(k) = \frac{(d/dt)(g(k) - kg'(k) - a)}{g(k) - kg'(k) - a} = \frac{-kg''(k)\dot{k}}{g(k) - kg'(k) - a}. \tag{24}$$

Using the definition of the elasticity of substitution the above will be reduced to

$$g(k)\sigma(k)\left[1 - \frac{a}{g(k) - kg'(k)}\right] = \dot{k}. \tag{25}$$

Solving the above differential equation (the Hotelling differential equation), we obtain

$$\int \frac{dk}{g(k)\sigma(k)[1 - a/(g(k) - kg'(k))]} = \int dt$$

or

$$\Psi(k; a) = t + C_0.$$

This solution specifies how the capital-exhaustible resource ratio, k, must follow the optimal path specified by Hotelling's rule. This equation together with the conservation law would allow the complete solutions for K and R.

Take the case of $\sigma(k) = 1$ (Cobb–Douglas). Then the Hotelling differential equation becomes

$$\frac{dk}{dt} = k^{\alpha} - \frac{a}{1-\alpha},$$

which will receive more detailed analysis in the next section.

6. An Example

Consider the case of the Cobb–Douglas production function. The Hotelling differential equation can be solved from

$$\int \frac{dk}{k^{\alpha} - a'} = \int dt, \; a' = \frac{a}{1-\alpha}. \tag{26}$$

If we assume that α is a rational number[1] $\alpha = m/n, n > m > 0$ (m and $n =$ integers), the above can be transformed to

$$\int \frac{dk}{k^{\alpha} - a'} = na'^{(n-m)/m} \int \frac{Z^{n-1}}{Z^m - 1} dZ, \tag{27}$$

where $Z = a'^{-1/m} k^{1/n}$. When $\alpha = 1/2$ or $m = 1$ and $n = 2$, we have a solution,

$$\int \frac{dk}{k^{1/2 - a'}} = 2a' \left(\frac{k^{1/2}}{a'} + \log \left(\frac{k^{1/2}}{a'} - 1 \right) \right). \tag{28a}$$

Unfortunately, when $a' \neq 0$, k can not be explicitly solved and, hence, we concentrate on the case of no extraction cost, i.e., $a = 0$. Then the Hotelling differential equation has a simple solution

$$\int \frac{dk}{k^{\alpha}} = \frac{1}{1-\alpha} k^{1-\alpha} = t + C_0 \tag{28b}$$

or

$$k = \frac{K}{R} = (1-\alpha)^{1/(1-\alpha)} (t + C_0)^{1/(1-\alpha)} \tag{29}$$

When $a = 0$, no extraction cost, our new policy rule (18) becomes

$$\dot{K} = f_R \left(R + \frac{A}{1 + q_2} \right). \tag{30}$$

[1] When α is not a rational number, it is known that no solution exists for Eq. (26), which can be expressible by combinations of elementary functions.

Using (29) to substitute R into (30), we obtain the following differential equation:

$$\dot{K} = \frac{K}{t + C_0} + b(t + C_0)^{\alpha/(1-\alpha)}, \tag{31}$$

where $b = (1-\alpha)^{1/(1-\alpha)} A/(1+q_2)$. To solve this differential equation, we introduce new variable, u,

$$u = \frac{K}{t + C_0}. \tag{32}$$

Then Eq. (31) becomes

$$\dot{u} = b(t + C_0)^{(2\alpha-1)/(1-\alpha)} \tag{33}$$

By integrating the above we get

$$u = b\frac{1-\alpha}{\alpha}(t + C_0)^{\alpha/(1-\alpha)} + C_1. \tag{34}$$

Since $u = K/(t + C_0)$, the time path of capital stock is

$$K = \frac{(1-\alpha)^{(2-\alpha)/(1-\alpha)}}{\alpha}\frac{A}{1+q_2}(t + C_0)^{1/(1-\alpha)} + C_1(t + C_0) \tag{35}$$

and time path of resource use becomes

$$R = C_1(1-\alpha)^{-1/(1-\alpha)}(t + C_0)^{-\alpha/(1-\alpha)} + \frac{1-\alpha}{\alpha}\frac{A}{1+q_2} \tag{36}$$

and time paths of q_1 and q_2 are given by

$$q_1 = C_2(t + C_0)^{-\alpha/(1-\alpha)}, \tag{37}$$

$$q_2 = C_2(1-\alpha)^{1/(1-\alpha)} - 1. \tag{38}$$

Also, we can derive consumption path from the budget constraint

$$\begin{aligned} C &= K^{\alpha}R^{1-\alpha} - \dot{K}, \\ C &= (1-\alpha)^{\alpha/(1-\alpha)}(t + C_0)^{\alpha/(1-\alpha)} \\ &\quad \times \left[C_1(1-\alpha)^{-1/(1-\alpha)}(t + C_0)^{-\alpha/(1-\alpha)} + \frac{1-\alpha}{\alpha}\frac{A}{1+q_2}\right] \\ &\quad - \frac{(1-\alpha)^{(2-\alpha)/(1-\alpha)}}{\alpha}\frac{A}{1+q_2}(1-\alpha)^{-1}(t + C_0)^{\alpha/(1-\alpha)} - C_1, \\ C &= \frac{\alpha}{1-\alpha}C_1. \end{aligned} \tag{39}$$

Therefore, $\dot{C} = 0$.

Even though we use Eq. (17) as a conservation law, the result is going to be the same as above. In fact when $a = 0$ through (19), Eq. (17) becomes

$$f_K K = C + \frac{A}{1+q_2} f_R. \tag{40}$$

By the assumption of Cobb–Douglas production function, the conservation law, Eq. (40) can be written as

$$\alpha K^{\alpha} R^{1-\alpha} = C + (1-\alpha)\frac{A}{1+q_2} K^{\alpha} R^{-\alpha}. \tag{41}$$

With Hotelling's rule, (28b) with solution (29), and Eq. (41), we can easily obtain the time path of capital stock, K,

$$K = \frac{(1-\alpha)^{(2-\alpha)/(1-\alpha)}}{\alpha} \frac{A}{1+q_2}(t+C_0)^{1/(1-\alpha)} + \frac{1-\alpha}{\alpha} C(t+C_0). \tag{42}$$

This time path of capital stock is exactly the same as Eq. (35) which is obtained from the previous conservation law, Eq. (16).

7. Benthamite Social Welfare Maximization

This section deals with a simple model for optimal capital accumulation with extraction costs of resources under the Benthamite utility function. We use the same assumptions as before, i.e., the constant-returns-to-scale technology (Eq. (2)) and a constant extraction cost (Eq. (11)). The economy has a representative agent with a well-defined utility function,

$$U = U(C(t)), \tag{43}$$

where $C(t)$ is consumption at time t.

The society's objective is to solve the Benthamite problem of dynamic optimization with a positive discount rate as follows:

$$\text{MAX} \int_0^{\infty} e^{-\rho t} U(C(t))dt \tag{44}$$

subject to

$$\dot{K} = f(K, R) - C - aR, \tag{45}$$

$$\dot{x} = -R(t). \tag{46}$$

Eq. (46) is an alternative representation of an iso-perimetric constraint

$$X = S - \int_0^t R(s)ds, \tag{47}$$

where S is total stock of exhaustible resources known at the beginning of time. By differencing the above equation with respect to time t, we can obtain Eq. (46). The Lagrange function for the above dynamic optimization problem can be written as

$$\mathcal{L} = e^{-\rho t}U(C) + \lambda_1[f(K,R) - \dot{K} - C - aR] + \lambda_2[\dot{X} + R] \\ = e^{-\rho t}V(\cdot), \tag{48}$$

where $V(\cdot) = U(C) + e^{\rho t}\lambda_1[f(K,R) - \dot{K} - C - aR] + e^{\rho t}\lambda_2[\dot{X} + R]$.

The Euler–Lagrange equations are

$$\mathcal{L}_C = 0, \quad e^{-\rho t}U' = \lambda_1, \tag{49a}$$

$$\mathcal{L}_K - \frac{d}{dt}\mathcal{L}_{\dot{K}} = 0, \quad \lambda_1 f_K + \dot{\lambda}_1 = 0, \tag{49b}$$

$$\mathcal{L}_R = 0, \quad \lambda_1(f_R - a) + \lambda_2 = 0, \tag{49c}$$

$$\frac{d}{dt}\mathcal{L}_{\dot{X}} = 0, \quad \dot{\lambda}_2 = 0. \tag{49d}$$

First, it is easily shown that Hotelling's rule can be derived for the above equations, by differentiating (49c) with respect to time and equating it to (49b) as

$$f_K = \frac{(d/dt)(f_R - a)}{f_R - a}, \quad f_R - a > 0.$$

Also λ_1 can be expressed as

$$\lambda_1 = \frac{\text{constant}}{f_R - a} = \frac{A}{f_R - a}$$

and U' is equal to

$$U' = \frac{e^{\rho t}A}{f_R - a}\cdot \tag{51}$$

We can now derive the investment function from the conservation law of the model. According to Sato (1990), there exists the income-wealth conservation law derived from

$$V(\cdot) - \dot{K}\frac{\partial V}{\partial \dot{K}} - \dot{X}\frac{\partial V}{\partial \dot{X}} = \rho \int_t^\infty e^{-\rho(s-t)}V(s)ds. \tag{52}$$

Applying (52) to this model we immediately obtain

$$\dot{K} = (f_R - a)R + \frac{f_R - a}{\lambda_2}\left[e^{-\rho t}U(t) - \rho \int_t^\infty e^{-\rho s}U(s)ds\right]. \tag{53}$$

For economic interpretation purpose the quantity $\int_t^\infty e^{-\rho s}U(s)ds$ is considered to be the discounted value of the total wealth measured in utility, while $e^{-\rho t}U(C)$ is the discounted value of utility. Thus,

$$e^{-\rho t}U(t) - \rho \int_t^\infty e^{-\rho s}U(s)ds \tag{54}$$

is the difference between the discounted consumption and the discounted total wealth multiplied by the discount rate ρ. If $\rho =$ interest rate, the second part of

Eq. (54) is looked as the total profits derived from the discounted wealth. Eq. (53) can alternatively be expressed by using the marginal utility of consumption.

$$\dot{K} = (f_R - a)R + \frac{1}{U'(C)}\left[\rho \int_t^{\infty} e^{-\rho s} U(s) ds - e^{-\rho t} U(t)\right]. \tag{55}$$

Also, by using the alternative formulation of the income-wealth conservation law, i.e., Net-Productivity Relation (Sato, 1990),

$$\frac{d}{dt}\left[V(\cdot) - \dot{K}\frac{\partial V}{\partial \dot{K}} - \dot{X}\frac{\partial V}{\partial \dot{X}}\right] = -\rho\left[\dot{K}\frac{\partial V}{\partial \dot{K}} + \dot{X}\frac{\partial V}{\partial \dot{X}}\right],$$

$$\frac{d}{dt}(\text{Utility Measure of Income}) = \rho(\text{Utility of Investment at } t) \tag{56}$$

We can express the investment rule. Using the Euler-Language equations we can write (56) as

$$\frac{dU}{dt} + \frac{d}{dt}\left[\lambda_1 e^{\rho t}(\dot{K} - (f_R - a)R)\right] = \rho\lambda_1 e^{\rho t}(\dot{K} - (f_R - a)R)$$

or

$$\frac{dU}{dt} + \rho\lambda_1 e^{\rho t}(\dot{K} - (f_R - a)R) + e^{\rho t}\dot{\lambda}_1(\dot{K} - (f_R - a)R)$$
$$+ e^{\rho t}\lambda_1 \frac{d}{dt}(\dot{K} - (f_R - a)R) = \rho\lambda_1 e^{\rho t}(\dot{K} - (f_R - a)R)$$

or

$$\dot{C} - \rho(\dot{K} - (f_R - a)R) + \frac{U''}{U'}\dot{C}(\dot{K} - (f_R - a)R) + \frac{d}{dt}(\dot{K} - (f_R - a)R) = 0.$$

Since $U' \neq 0$ and

$$\frac{U''\dot{C}}{U'} = \left(\rho - \frac{(d/dt)(f_R - a)}{f_R - a}\right)$$

or

$$\frac{U''\dot{C}}{U'} = \rho - f_K, \tag{57}$$

we finally obtain

$$\dot{K} = (f_R - a)R + \frac{(d/dt) f_K K}{f_K}. \tag{58}$$

The above is the alternative formulation of the investment function, alternative to Eq. (53). It tells us that investment also depends on how profit ($f_K K$) changes over time.

It is interesting to note that with the discount rate $\rho \neq 0$, consumption will not be constant. From (57) we know that

$$\dot{C} > 0 \text{ if } f_K > \rho,$$

$$\dot{C} = 0 \text{ if } f_K = \rho,$$

$$\dot{C} < 0 \text{ if } f_K < \rho.$$

As, in the long run, the marginal product of K tends to decrease as the exhaustible resource becomes more and more scarce, f_K tends to be less than ρ. This means that, from the Euler–Lagrange equations, optimal consumption is constant for all t if and only if $\rho \equiv 0$. That is to say, Solow's Max–Min principle applies to the Benthamite problem if and only if there is no discount rate, a well-known result (Solow, 1974).

8. Concluding Remarks

In the existing studies on the intergenerational equity problem of optimal capital accumulation and exhaustible resource extraction, an investment rule, Hartwick's rule, has been proposed. The rule asks to keep the total value of net investment equal to zero. In this paper, we use conservation laws to investigate investment rules for intergenerational equity and Benthamite utility maximization, respectively. From a simple conservation law based on the Max–Min principle for intergenerational equity, we obtain the generalized investment rule indicating the value of net investment equal to a constant, which is sufficient for constant consumption. However, in the Benthamite utility maximization the investment rule as a conservation law is the summation of the discounted difference in value of utility from total wealth and the investment rule from the Max–Min principle. The investment rule in the Benthamite utility maximization is reduced to the rule from the Max–Min principle whenever the discount rate approaches zero. This means that Solow's Max–Min principle applies to the Benthamite problem if and only if there is no time preference.

References

Dixit, A., Hammond, P., Hoel, M., 1980. On Hartwick's rule for regular maximin paths of capital accumulation and resource depletion. Review of Economic Studies 47, 551–556.

Hartwick, J., 1977. Intergenerational equity and the investing of rents from exhaustible resources. American Economic Review 66, 972–974.

Sato, R., 1990. In: Sato, R., Ramachandran, R. (Eds.), The Invariance Principle and Income-Wealth Conservation Laws, Conservation Laws and Symmetry: Applications to Economics and Finance. Kluwer, Dordrecht, pp. 71–106.

Solow, R., 1974. Intergenerational equity and exhaustible resources, review of economic studies. Symposium on the Economics of Exhaustible Resources, pp. 29–45.

Solow, R.M., Wan, F.Y., 1975. Extraction costs in the theory of exhaustible resources. The Bell Journal of Economics, 7 (2), 359–370.

Appendix

DYNAMIC SYMMETRIES AND ECONOMIC CONSERVATION LAWS

1. Introduction

Motivation: Dynamic Symmetries in Natural Systems

1. One of the most fascinating aspects of the group-theoretic method in the different branches of physical sciences is the analysis of the connection between dynamic symmetries (invariances) and conservation laws of a mathematical system. When natural systems are idealized, unexpected symmetries tend to arise. These symmetries are not apparent at first and are only revealed through rigorous analysis. The past 50 years have exhibited remarkable development in this new area of mathematics, which has far-reaching significance in the fields of physics, engineering, and applied mathematics.[1]

The fundamental work on this problem was initiated in the early part of this century by Noether [1918] in a classic paper.[2] Influenced by the work of Klein [1918] and of Lie [1891] on the transformation properties of differential equations under continuous (Lie) groups, Noether proved the fundamental result, now known as the *Noether theorem*. Her work represents an ingeneous combination of the methods of the formal calculus of variations with those of Lie's group theory.

Noether's theorem has important consequences. For example, if L is the Lagrangian of some physical or economic system, then the invariance of

[1] See Klein [1918], Lie [1891], Noether [1918], Bessel-Hagen [1921], Gelfand and Fomin [1963], Goldstein [1950], Rund [1966], Logan [1977], Nôno [1968], Nôno and Mimura [1975, 1976, 1977], Sagan [1969], Lovelock and Rund [1975], Lanczos [1970], Whittaker [1937], and Moser [1979].

[2] Her paper has recently been translated into English by Tavel, who also supplies a brief motivation and historical sketch (see Noether [1918]).

the fundamental integral under an *r-parameter (Lie) group leads directly to conservation laws for the system.* This means that the invariance hypothesis leads to expressions which are *constant* along the extremals, or first integrals of the equations of motion. The Noether theorem was utilized by Bessel-Hagen [1921] to compute the conservation laws in particle mechanics, using the ten-parameter Galilean group, which consists of a time translation, spatial translations and rotations, and velocity translations. The resulting ten conservation laws are the classical ten integrals of the N-body problem: conservation of energy, linear momentum, angular momentum, and the uniform motion of the center of mass. A group invariance such as a rotation or translation invariance is often referred to as a *symmetry* of a dynamic system or simply a *dynamic symmetry*.[3]

Economic Conservation Laws

2. In his well-known article, Ramsey [1928, p. 547] states his famous rule of optimal saving as "rate of saving multiplied by marginal utility of consumption should always equal bliss minus actual rate of utility enjoyed." This rule was derived from

$$\dot{c} = \frac{dc}{dt} = f(a,\ c) - x = \frac{B - (U(x) - V(a))}{u(x)},$$

where f is the production function, a the labor, c the capital, x the consumption, $U(x)$ the utility of consumption, $V(a)$ the disutility of labor, and $u(x) = \partial U/\partial x$ the marginal utility of consumption. Although Ramsey did not mention it explicitly, his rule is closely related to and is derived from *the law of conservation of energy*. In a conservative dynamic system, the Hamiltonian defined by the Legendre transformation

$$H = \sum_{i=1}^{n} p_i \dot{x}_i - L \quad (p_i = \partial L/\partial \dot{x}_i,\ L = \text{Lagrangian})$$

is *constant*. By letting $L = -(U(x) - V(a))$ and Bliss $B = H$, the Ramsey rule is derived from the conservation law

$$H = B = u(x)\dot{c} + (U(x) - V(a)) = u(x)\dot{c} - L.$$

Samuelson [1970a, b] is the first economist *explicitly* to introduce conservation laws in theoretical economics. From the analogy of the law of conservation of (physical) energy, kinetic energy + potential energy = constant, Samuelson obtained the conservation law of the aggregate capital–output ratio in a

[3] It may happen, however, that these symmetries are not at all obvious but quite opaque. A famous example is the Kepler problem which asks for the motion of a single particle under the influence of a fixed mass point (say, the sun) according to Newton's inverse-square law of gravitation. The solutions are ellipses, hyperboles, or parabolas. This system is rotation symmetric, and therefore has a fixed angular momentum vector. But it turns out—and this is by no means obvious—that the Kepler problem admits not only the rotation group in three-dimensional space, but even in four-dimensional space. This additional symmetry was fully understood first for the quantum-theoretical Kepler problem (Moser [1979]).

neoclassical von Neumann economy, where all output is saved to provide capital formation for the growth of the system. Although he did not appeal explicitly to Noether's theorem, it is seen that Samuelson's conservation laws are the most fundamental laws for the closed, consumptionless system of the von Neumann type. (This will be proved in section 18 using the Noether theorem directly.)

As a basis of motivation and in order to show the apparent connection between the (static) holotheticity concept in earlier chapters and the dynamic invariance (or dynamic symmetry, or dynamic holotheticity) concept in this chapter, let us consider the problem posed by Samuelson [1970a, b]. The optimal-control problem is defined as

$$\max_{\{k^i(t)\}} \int_0^T \dot{K}^1(t)\, dt \quad \text{subject to } F(K(t),\ \dot{K}(t)) = 0$$

and assigned boundary conditions, where F is a smooth, neoclassical, concave, first-degree homogeneous transformation function relating the vector of n capital goods $K(t) = \{K^i(t)\}$ and the vector of net capital formation $\dot{K} = \{dK^i/dt\}$. This variational problem is to maximize the functional

$$\begin{aligned} J(K(t)) &= \int_0^T L(K(t),\ \dot{K}(t),\ \lambda(t))\, dt \\ &= \int_0^T (\dot{K}^1(t) + \lambda(t)F[K(t),\ \dot{K}(t)])\, dt. \end{aligned}$$

Let us suppose that we are given the time translation $\bar{t} = t + \varepsilon$ with no technical progress for capital goods, $\bar{K}^i = K^i$ and $\bar{\lambda} = \lambda$. We are interested in finding the condition that the fundamental integral is invariant under the time translation, i.e.,

$$\int_c L(\bar{K}(\bar{t}),\ \dot{\bar{K}}(\bar{t}),\ \bar{\lambda}(\bar{t}))d\bar{t} = \int_c L(K(t),\ \dot{K}(t), \lambda(t))\, dt.$$

In order that this requirement be satisfied, it is necessary and sufficient that $L(\bar{K}(\bar{t}), \dot{\bar{K}}(\bar{t}), \bar{\lambda}(\bar{t}))\, dt/dt = L(K(t),\ \dot{K}(t), \lambda(t))$. But, since $d\bar{t}/dt = 1$, this requirement is nothing but

$$L(\bar{K}(\bar{t}),\ \dot{\bar{K}}(\bar{t}), \bar{\lambda}(\bar{t})) = L(K(t), \dot{K}(t), \lambda(t)).$$

This condition is simply the (static) holotheticity condition, when the differential equation L does *not* contain the independent variable t explicitly. (This corresponds to the special case of the infinitesimal transformation $U = \partial/\partial t$, discussed as the additive case in Chapter 2.) Hence, in this case, the (*static*) *holotheticity condition is exactly identical with the dynamic invariance condition.*

In general, however, since L contains time t and also since there exists technical change for capital goods, the static condition is *not* equal to the dynamic invariance. We shall later present more precise definitions of dynamic invariance conditions.

3. The purpose of this chapter is to present a self-contained and systematic analysis of conservation laws in economic dynamics. First, a fairly detailed

mathematical treatment of Noether's theorem will be given, so that the economist who is not familiar with this aspect of variational dynamics can learn its applications. Second, the invariant variational principle will be applied to *general* neoclassical optimal growth models of the Ramsey type. It is shown that there exist several (local) conservation laws operating in the neighborhood of steady positions. An attempt will be made to derive more general local conservation laws that operate in a variety of special dynamic systems. Finally, the Samuelson conservation laws in a von Neumann growth system will be derived through the application of the Noether theorem. It will be shown that the Samuelson conservation laws are the only laws operating globally for the system.

2. Preliminaries: Noether's Theorem and Invariance Identities[4]

r-Parameter Transformations: Technical Change with *r* Essential Parameters

4. To motivate interest and to introduce the idea of (dynamic) invariance of the *fundamental* or *action integral* of a given calculus of variations or optimal control problem, we consider a classical result concerning an integration of the Euler–Lagrange equations. As in the case of Samuelson's problem cited in Section 2, if the Lagrangian L does not depend explicitly on t,

$$J(x) = \int_{t_0}^{t_1} L(x(t),\ \dot{x}(t))\, dt,\ x(t) = (x^1(t),\ \ldots,\ x^n(t)),$$
$$\dot{x}(t) = (\dot{x}^1(t),\ \ldots,\ \dot{x}^n(t)),$$

then the Euler–Lagrange equations have first integral

$$-L + \dot{x}^k \frac{\partial L}{\partial \dot{x}^k} = H = C \quad \left(\dot{x}^k \frac{\partial L}{\partial \dot{x}^k} = \sum_{k=1}^{n} \dot{x}^k \frac{\partial L}{\partial \dot{x}^k},\ \text{summation convention}^5 \right),$$

where H is the Hamiltonian and C is a constant (see Whittaker [1937]). To say that L does not depend explicitly on t could be interpreted as saying that the functional J is invariant under the transformation which takes t to $t + \varepsilon$. The Noether theorem can to a certain degree be considered as a generalization of this fact. Let us consider a Lagrange function L: $I \times \mathbb{R}^n \times \mathbb{R}^n \to \mathbb{R}^1$, which is twice continuously differentiable in each of its $2n + 1$ arguments, $L \in C^2$, and $I \subseteq R^1$ is an open interval of real numbers. We then have the variational integral

$$J(x) = \int_a^b L(t,\ x(t),\ \dot{x}(t))\, dt, \tag{1}$$

[4] The reader who is familiar with Noether's theorem and its applications can skip this part and proceed to Part III.

[5] We adopt the so-called Einstein summation convention (Young [1978, pp. 333–334]. When a lowercase Latin index such as j, h, k, ... appears twice in a term, summation over that index is implied, the range of summation being 1, ..., n. For example, $a^i x^i$ means $\sum_{i=1}^n a^i x^i$.

where $[a, b] \subseteq 1$ and $x \in C_n^2[a, b]$ is the set of all vector functions $x(t) = (x^1(t), \ldots, x^n(t))$, $t \in [a, b]$. The type of invariance transformations that will be considered are (technical change) transformations of configuration space, i.e., $(t, x^1, \ldots, x^n)$ space, which depend on r real, independent (essential) parameters $\varepsilon^1, \ldots, \varepsilon^r$. In Noether's original paper [1918], as well as in more recent treatments of invariance problems, it is assumed that the transformations form a group; in fact, the groups considered are precisely local *Lie groups* of transformations. Although many of the transformations which arise in applications to mathematical physics and mathematical economics are group of transformations (e.g., the Galilean group, the Lorenz group, biased technical change), the group concept is not necessarily required. To be more precise, we require here that the transformations are given by

$$\begin{aligned} \bar{t} &= \phi(t, x, \varepsilon), \quad \varepsilon = (\varepsilon^1, \ldots, \varepsilon^r), \\ \bar{x}^k &= \psi^k(t, x, \varepsilon) \quad (k = 1, \ldots, n). \end{aligned} \tag{2}$$

In economic terms, ϕ can be considered as "subjective" time (Samuelson [1976]), while ψ^k represent "technical" or "taste" change. It is assumed that a domain U in $\mathbb{R}^r$ is given which contains the origin as an interior point and that the mappings

$$\phi\colon [a,b] \times \mathbb{R}^n \times U \to I, \quad \psi^k\colon [a, b] \times \mathbb{R}^n \times U \to I \quad (k = 1, \ldots, n)$$

are of class C^2 in each of their $1 + n + r$ arguments with

$$\phi(t, x, 0) = t, \quad \psi^k(t, x, 0) = x^k \quad (k = 1, \ldots, n). \tag{3}$$

Expanding the right-hand sides of (2) in a Taylor series about $\varepsilon = 0$, we obtain

$$\begin{aligned} \bar{t} &= t + \tau_s(t, x)\varepsilon^s + O(\varepsilon), \\ \bar{x}^k &= x^k + \xi_s^k(t, x)\varepsilon^s + O(\varepsilon) \quad (k = 1, \ldots, n) \end{aligned} \tag{4}$$

($s = 1, \ldots, r$, summation convention in force). The principal linear parts τ_s and ξ_s^k are nothing but the *infinitesimal generators* (or *transformations*) of the transformation (2) given by

$$\tau_s(t, x) = \frac{\partial \phi}{\partial \varepsilon^s}(t, x, 0), \quad \xi_s^k(t, x) = \frac{\partial \psi^k}{\partial \varepsilon^s}(t, x, 0). \tag{5}$$

Let $x\colon [a, b] \to \mathbb{R}^n$ be a curve of class C_n^2 given by $x = x(t)$; then, for sufficiently small ε, the transformation

$$\bar{t} = \phi(t, x(t), \varepsilon) \tag{6}$$

is invertible (see Logan [1977]).[6] Writing the unique inverse function of (6) as $t = T(\bar{t}, \varepsilon)$, and substituting into the equation $\bar{x}^k = \psi^k(t, x(t), \varepsilon)$, we get

$$\begin{aligned} \bar{x}^k &= \psi^k(T(\bar{t}, \varepsilon), x(T(\bar{t}, \varepsilon)), \varepsilon) \\ &\equiv \bar{x}^k(\bar{t}) = \bar{x}^k(\phi(t, x(t), \varepsilon)) = \psi^k(t, x(t), \varepsilon). \end{aligned} \tag{7}$$

[6] If we assume a local Lie group transformation, then by definition it is invertible and has the same inverse form as the original mapping.

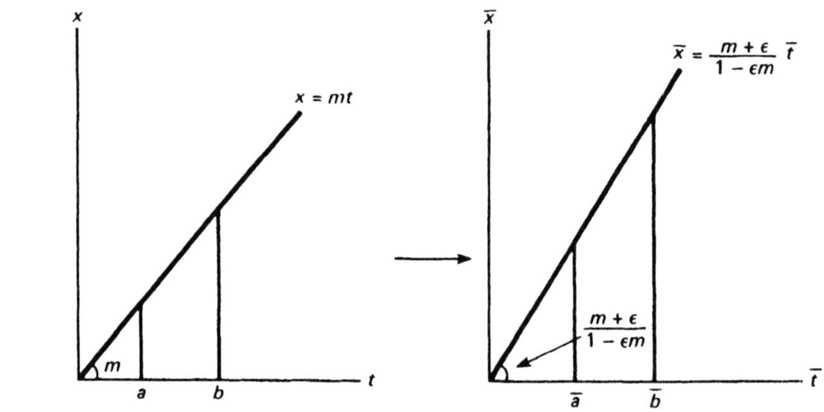

FIGURE 1. Rotation Group

Hence we may define the transformed curves $\bar{x} = \bar{x}(\bar{t})$ in $(\bar{t},\ \bar{x})$ space. By setting

$$\bar{a} = \phi(a,\ x(a),\ \varepsilon), \quad \bar{b} = \phi(b,\ x(b),\ \varepsilon), \tag{8}$$

$\bar{x}(\bar{t})$ is defined on $[\bar{a},\ \bar{b}]$. As an example, consider a simple rotation group $(n = 1,\ r = 1)$ $\bar{t} = t - \varepsilon x,\ \bar{x} = x + \varepsilon t$. Here the infinitesimal transformation is given by $U = -x\partial/\partial t + t\ \partial/\partial x$. Consider the curve $x(t) = mt,\ m = \text{const}$. We then have $\bar{t} = t - \varepsilon mt,\ \bar{x} = mt + \varepsilon t$. Solving the first equation for t in terms of $\bar{t}$, we get $t = \bar{t}/(1 - \varepsilon m),\ 1 - \varepsilon m \neq 0$. Thus $\bar{x} = ((m + \varepsilon)/(1 - \varepsilon m))\bar{t} \equiv \bar{x}(\bar{t},\ \varepsilon)$. If t is defined for $[a,\ b]$, then $\bar{a} = (1 - \varepsilon m)a$ and $\bar{b} = (1 - \varepsilon m)b$ and $\bar{t}$ is defined for $[\bar{a},\ \bar{b}]$. Figure 1 depicts this transformation.

Invariance Definitions

5. We first present a geometrical interpretation of the transformation for a general case of $n = 1$ and $r = 1$. Given $x = x(t)$ in the tx plane, then the curve $x = x(t)$ is mapped to a one-parameter family of curves under the transformation $t \to \bar{t} = \phi(t,\ x(t),\ \varepsilon)$ and $x \to \bar{x} = \psi(t,\ x(t),\ \varepsilon)$. The resulting family of curves are the graphs of functions in the $\bar{t}\bar{x}$ plane, provided ε is sufficiently small. Figure 2 depicts this transformation with the curve drawn in the $\bar{t}\bar{x}$ plane showing one curve out of the one-parameter family for a particular value of ε. In the tx plane we can calculate the functional

$$J(x) = \int_a^b L(t,\, x(t),\, \dot{x}(t))\, dt, \quad \dot{x}(t) = dx(t)/dt.$$

Also, for the $\bar{t}\bar{x}$ plane we can calculate

$$J(\bar{x}) = \int_{\bar{a}}^{\bar{b}} L(\bar{t},\ \bar{x}(\bar{t}),\ \dot{\bar{x}}(\bar{t}))\, d\bar{t}, \quad \dot{\bar{x}}(\bar{t}) = d\bar{x}(\bar{t})/d\bar{t}.$$

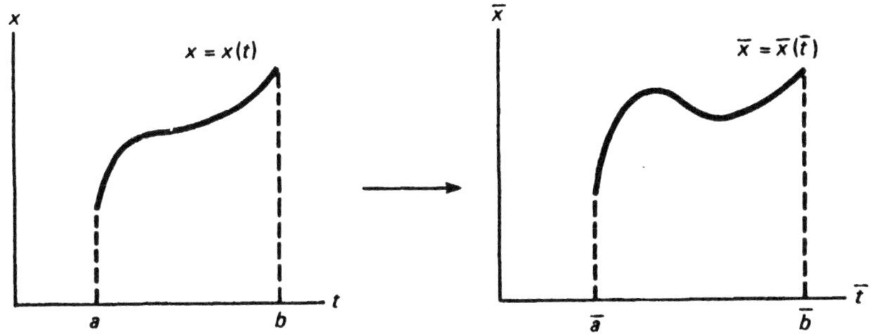

FIGURE 2. One-Parameter Transformation

We say that J is invariant under the given transformation if $J(x) = J(\bar{x})$ or at least so up to the first-order terms in ε.

Definition 1 (*Dynamic Invariance or Dynamic Symmetry*)[7] (a) The fundamental integral (1) is absolutely invariant under the r-parameter family of transformations (2) if and only if given any curve $x\colon [a, b] \to R^n$ of class C_n^2 and $a \leqq t_1 < t_2 \leqq b$, we have

$$\int_{\bar{t}_1}^{\bar{t}_2} L\left(\bar{t},\ \bar{x}(\bar{t}),\ \frac{d\bar{x}(\bar{t})}{d\bar{t}}\right) d\bar{t} - \int_{t_1}^{t_2} L\left(t,\ x(t),\ \frac{dx(t)}{dt}\right) dt = O(\varepsilon) \tag{9}$$

for all $\varepsilon \in U$ with $|\varepsilon| < d$, where d depends on $x(t)$, where $\bar{x}(\bar{t})$ is defined by (7), and where $\bar{t}_i = \phi(t_i, x(t_i), \varepsilon)$, $i = 1, 2$.

(b) Alternatively, the fundamental integral is absolutely invariant if and only if

$$L\left(\bar{t}, \bar{x}(\bar{t}),\ \frac{d\bar{x}(\bar{t})}{d\bar{t}}\right) \frac{d\bar{t}}{dt} - L\left(t,\ x(t),\ \frac{dx(t)}{dt}\right) = O(\varepsilon). \tag{10}$$

By transforming the first integral in (9) back to the interval $[t_1, t_2]$, condition (9) is entirely equivalent to (10), which involves only the Lagrangian and not the action integral.

Remark 1 Definition 1 is somewhat more general than the one usually given (see Gelfand and Fomin [1963], Sagan [1969], Rund [1966], Lovelock and Rund [1975]), since $J(\bar{x}) - J(x)$ is not necessarily zero, but is equal to first-order terms in ε. If $J(\bar{x}) - J(x) = 0$, we can show that

$$J(\varepsilon) = \int_{\bar{t}_1(\varepsilon)}^{\bar{t}_2(\varepsilon)} L\left(\bar{t},\ \bar{x}(\bar{t}, \varepsilon),\ \frac{d\bar{x}(\bar{t},\ \varepsilon)}{d\bar{t}}\right) d\bar{t}$$

is independent of ε; i.e., $dJ(\varepsilon)/d\varepsilon = 0$ for all ε.

Remark 2 The invariance definition obviously holds true regardless of whether the Euler–Lagrange equations vanish or not.

Example 1 Consider $L = T - U$, where T is the kinetic energy given by $T = \frac{1}{2} m_k \dot{x}^k$, and U is the potential energy $U(t, x)$. Suppose that the potential energy is translation invariant, i.e., $U(t,\ \bar{x}) = U(t,\ x)$ where $\bar{x}^k = x^k + \varepsilon$, $\bar{t} = t$.

[7] In contrast to cases in which a function (rather than functional) is invariant, which was the main topic up to this chapter, we may call dynamic invariance as "dynamic holotheticity."

From

$$\frac{d\bar{t}}{dt} = 1 \quad \text{and} \quad \frac{d\bar{x}^k}{d\bar{t}} = \frac{d\bar{x}^k}{dt} \bigg/ \frac{d\bar{t}}{dt} = \frac{dx^k}{dt},$$

condition (10) requires that

$$\begin{aligned}&\frac{1}{2}m_k\dot{\bar{x}}^k - U(t,\ \bar{x}) - (\frac{1}{2}m_k\dot{x}^k - U(t,\ x))\\ &= \frac{1}{2}m_k\dot{\bar{x}}^k - \frac{1}{2}m_k\dot{x}^k - U(t,\ \bar{x}) + U(t,\ x) \equiv 0.\end{aligned}$$

In this case we have $L(\bar{t},\ \bar{x},\ \dot{\bar{x}}) - L(t, x, \dot{x}) \equiv 0$.

Example 2 Let $J(x)$ be given by

$$J(x) = \int_a^b \sqrt{1 + \dot{x}(t)^2}dt$$

and the one-parameter group of rotation by $\bar{t} = t - \varepsilon x,\ \bar{x} = x + \varepsilon t$. From $d\bar{t}/dt = 1 - \varepsilon\dot{x}$ and $d\bar{x}/dt = \dot{x} + \varepsilon$, we have

$$\frac{d\bar{x}}{d\bar{t}} = \frac{d\bar{x}}{dt} \bigg/ \frac{d\bar{t}}{dt} = \frac{\dot{x} + \varepsilon}{1 - \varepsilon\dot{x}}.$$

Therefore, condition (10) requires

$$\begin{aligned}\left(\sqrt{1 + \left(\frac{d\bar{x}}{d\bar{t}}\right)^{-2}}\right)\frac{d\bar{t}}{dt} - \sqrt{1 + \dot{x}^2} &= \left(\sqrt{1 + \left(\frac{\dot{x} + \varepsilon}{1 - \varepsilon\dot{x}}\right)^{-2}}\right)(1 - \varepsilon\dot{x}) - \sqrt{1 + \dot{x}^2}\\ &= \sqrt{(1 + \dot{x}^2)(1 + \varepsilon^2)} - \sqrt{1 + \dot{x}^2}\\ &= \sqrt{1 + \dot{x}^2}(\sqrt{1 + \varepsilon^2} - 1)\\ &= \sqrt{1 + \dot{x}^2}(1 + \frac{1}{2}\varepsilon^2 - \frac{1}{8}\varepsilon^4 + \ldots - 1)\\ &= \sqrt{1 + \dot{x}^2}(\frac{1}{2}\varepsilon^2 - \frac{1}{8}\varepsilon^4 + \ldots)\\ &= O(\varepsilon).\end{aligned}$$

Hence, according to Definition 1, the functional J in this case is also absolutely invariant under the rotation group.

Definition 2 (*Divergence-Invariant or Invariance with Nullity*) The fundamental integral is divergence-invariant, or invariant up to a divergence term, if there exist r functions $\Phi_s: I \times \mathbb{R}^n \to \mathbb{R}^1,\ s = 1,\ \ldots,\ r$, of class C^1 such that

$$L\left(\bar{t}, \bar{x}(\bar{t}), \frac{d\bar{x}(\bar{t})}{d\bar{t}}\right)\frac{d\bar{t}}{dt} - L(t,\ x(t),\ \dot{x}(t)) = \varepsilon^s(d\Phi_s(t,\ x(t))/dt + O(\varepsilon)$$

(summation convention on s), **(11)**

with the remaining conditions of Definition 1 holding true.

This definition is more general than Definition 1 in that the left-hand side of (10) is equal to the sum of the exact differentials multiplied by the essential parameters and to first-order terms in ε. Since the addition of the total exact differentials does *not* change the original Euler–Lagrange equations associated with L (see Sagan [1969]), we may call this case the *dynamic symmetry condition with nullity*, or simply the *dynamic invariance with nullity*. Alternatively, Definition 2 is referred to as "invariance up to an exact differential" (see Rund [1966, p. 73]).

The Fundamental Invariance Identities, the Noether Theorem, and Conservation Laws

6. If the fundamental integral is invariant (either absolutely or up to an exact differential) under an r-parameter family (or group) of continuous transformations, the corresponding Lagrangian must satisfy certain conditions involving the Lagrangian, its derivatives, and the infinitesimal generators of the continuous transformations. On using the Euler–Lagrange expressions, the resulting formulation of these conditions is usually referred to as Noether's theorem. The importance of this theorem lies in the fact that it allows one to construct quantities which are constant along any extremal, i.e., a curve which satisfies the Euler–Lagrange equations. Hence it allows one to obtain relations which may be interpreted as "conservation laws."

We first derive the so-called *invariance* identities. By differentiating (11) with respect to ε^s and then setting $\varepsilon = 0$, we obtain

$$\left\{\frac{\partial L}{\partial t}\tau_s + \frac{\partial L}{\partial x^k}\xi_s^k + \frac{\partial L}{\partial \dot{x}^k}\left(\frac{\partial}{\partial \varepsilon^s}\left(\frac{d\bar{x}^k}{d\bar{t}}\right)\right)_0\right\}\left(\frac{d\bar{t}}{dt}\right)_0 + L\left(\frac{\partial}{\partial \varepsilon^s}\left(\frac{d\bar{t}}{dt}\right)\right)_0 = \frac{d\Phi_s}{dt} \quad \text{(summation on } k = 1, \ldots, n), \tag{12}$$

where $(\cdot)_0$ denotes $(\cdot)_{\varepsilon=0}$. In order to be able to exploit this condition, we must evaluate the various derivatives with respect to ε^s which appear on the left-hand side. We need the following expressions which are obvious from (2) and (3):

$$\begin{aligned}
&\left(\frac{\partial \bar{t}}{\partial t}\right)_0 = 1, \quad \left(\frac{\partial \bar{x}^k}{\partial t}\right)_0 = \left(\frac{\partial \bar{t}}{\partial x^k}\right)_0 = 0,\\
&\left(\frac{\partial \bar{x}^k}{\partial x^h}\right)_0 = \delta_h^k \quad \text{(the Kronecker } \delta\text{)},\\
&\left(\frac{\partial^2 \bar{x}^k}{\partial \varepsilon^s \partial t}\right)_0 = \frac{\partial \xi_s^k}{\partial_t}, \quad \left(\frac{\partial^2 \bar{x}^k}{\partial \varepsilon^s \partial x^h}\right)_0 = \frac{\partial \xi_s^k}{\partial x^h},\\
&\left(\frac{\partial^2 \bar{t}}{\partial \varepsilon^s \partial t}\right)_0 = \frac{\partial \tau_s}{\partial t}, \quad \left(\frac{\partial^2 \bar{t}}{\partial \varepsilon^s \partial x^h}\right)_0 = \frac{\partial \tau_s}{\partial x^h}.
\end{aligned} \tag{13}$$

We observe that, along any curve, we have by the chain rule

$$\frac{d\bar{t}}{dt} = \frac{\partial \bar{t}}{\partial t} + \frac{\partial \bar{t}}{\partial x^k}\dot{x}^k.$$

Differentiating this equation with respect to ε^s, and setting $\varepsilon = 0$ and using (13), we get

$$\left(\frac{\partial}{\partial\varepsilon^s}\left(\frac{d\bar{t}}{dt}\right)\right)_0 = \frac{\partial\tau_s}{\partial t} + \frac{\partial\tau_s}{\partial x^k}\dot{x}^k = \frac{d\tau_s}{dt}. \tag{14}$$

To calculate other expressions $((\partial/\partial\varepsilon^s)(\partial\bar{x}^k/\partial t))_0$, we differentiate (7) with respect to t to obtain

$$\frac{\partial\bar{x}^k}{\partial t} + \frac{\partial\bar{x}^k}{\partial x^h}\dot{x}^h = \frac{d\bar{x}^k}{d\bar{t}}\frac{d\bar{t}}{dt} = \frac{d\bar{x}^k}{d\bar{t}}\left(\frac{\partial\bar{t}}{\partial t} + \frac{\partial\bar{t}}{\partial x^h}\dot{x}^h\right). \tag{15}$$

Setting $\varepsilon = 0$ and using (13), we get

$$\delta^k_h\dot{x}^h = (d\bar{x}^k/d\bar{t})_0(1 + 0\cdot\dot{x}^h);$$

i.e.,

$$\dot{x}^k = (d\bar{x}^k/d\bar{t})_0.$$

Again, differentiating (15) with respect to ε^s, we derive

$$\frac{\partial^2\bar{x}^k}{\partial\varepsilon^s\partial t} + \frac{\partial^2\bar{x}^k}{\partial\varepsilon^s\partial x^h}\dot{x}^h = \frac{d\bar{x}^k}{d\bar{t}}\left(\frac{\partial^2\bar{t}}{\partial\varepsilon^s\partial t} + \frac{\partial^2\bar{t}}{\partial\varepsilon^s\partial x^h}\dot{x}^h\right) + \left(\frac{\partial}{\partial\varepsilon^s}\left(\frac{\partial\bar{x}^k}{\partial\bar{t}}\right)\right)\left(\frac{\partial\bar{t}}{\partial t} + \frac{\partial\bar{t}}{\partial x^h}\dot{x}^h\right),$$

after which we set $\varepsilon = 0$ while using (13). This yields

$$\frac{\partial\xi^k_s}{\partial t} + \frac{\partial\xi^k_s}{\partial x^h}\dot{x}^h = \dot{x}^k\left(\frac{\partial\tau_s}{\partial t} + \frac{\partial\tau_s}{\partial x^h}\dot{x}^h\right) + \left(\frac{\partial}{\partial\varepsilon^s}\left(\frac{\partial\bar{x}^k}{\partial\bar{t}}\right)\right)_0,$$

since

$$\left(\frac{\partial\bar{t}}{\partial t} + \frac{\partial\bar{t}}{\partial x^h}\dot{x}^h\right)_0 = \left(\frac{d\bar{t}}{dt}\right)_0 = 1.$$

Hence we obtain

$$\left(\frac{\partial}{\partial\varepsilon^s}\left(\frac{\partial\bar{x}^k}{\partial\bar{t}}\right)\right)_0 = \frac{d\xi^k_s}{dt} - \dot{x}^k\frac{d\tau_s}{dt}. \tag{16}$$

Substituting (14) and (16) into (12), we derive the invariance identities

$$\frac{\partial L}{\partial t}\tau_s + \frac{\partial L}{\partial x^k}\xi^k_s + \frac{\partial L}{\partial\dot{x}^k}\left(\frac{d\xi^k_s}{dt} - \dot{x}^k\frac{d\tau_s}{dt}\right) + L\frac{d\tau_s}{dt} = \frac{d\boldsymbol{\Phi}_s}{dt} \tag{17}$$

($s = 1, \ldots, r$) (summation convention operative for $k = 1, \ldots, n$). The above are often referred to as the *fundamental invariance identities*, which involve the Lagrangian, its derivatives, and the infinitesimal generators τ_s and ξ^k_s of the transformations.

Noether's Theorem and Conservation Laws

7. The classical theorem of Noether on invariant variational problems can now be derived from the fundamental invariance identities when the Euler-Lagrange equations are introduced into this relation. Let the *Euler–Lagrange expressions* be given by

$$E_k = \frac{\partial L}{\partial x^k} - \frac{d}{dt}\frac{\partial L}{\partial \dot{x}^k} \quad (k = 1, \ldots, n). \tag{18}$$

We note that

$$\frac{\partial L}{\partial t}\tau_s = \left(\frac{dL}{dt} - \frac{\partial L}{\partial x^k}\dot{x}^k - \frac{\partial L}{\partial \dot{x}^k}\ddot{x}^k\right)\tau_s$$

and

$$\frac{\partial L}{\partial \dot{x}^k}\frac{d\xi_s^k}{dt} = \frac{d}{dt}\left(\frac{\partial L}{\partial \dot{x}^k}\xi_s^k\right) - \frac{d}{dt}\left(\frac{\partial L}{\partial \dot{x}^k}\right)\xi_s^k,$$

together with

$$\frac{\partial L}{\partial \dot{x}^k}\dot{x}^k\frac{d\tau_s}{dt} + \frac{\partial L}{\partial \dot{x}^k}\ddot{x}^k\tau_s = \frac{d}{dt}\left(\frac{\partial L}{\partial \dot{x}^k}\dot{x}^k\tau_s\right) - \frac{d}{dt}\left(\frac{\partial L}{\partial \dot{x}^k}\right)\dot{x}^k\tau_s.$$

These three relations are substituted in turn into (17) to yield

$$\begin{aligned}&\left[\frac{\partial L}{\partial x^k} - \frac{d}{dt}\left(\frac{\partial L}{\partial \dot{x}^k}\right)\right]\xi_s^k - \left[\frac{\partial L}{\partial x^k} - \frac{d}{dt}\left(\frac{\partial L}{\partial \dot{x}^k}\right)\right]\dot{x}^k\tau_s \\ &\quad + \frac{d}{dt}\left[L\tau_s + \frac{\partial L}{\partial \dot{x}^k}\xi_s^k - \frac{\partial L}{\partial \dot{x}^k}\dot{x}^k\tau_s\right] = \frac{d\Phi_s}{dt} \quad (s = 1, \ldots, r).\end{aligned}$$

Using (18), this can be expressed as

$$E_k(\xi_s^k - \dot{x}^k\tau_s) + \frac{d}{dt}\left[L\tau_s + \frac{\partial L}{\partial \dot{x}^k}\xi_s^k - \frac{\partial L}{\partial \dot{x}^k}\dot{x}^k\tau_s - \Phi_s\right] = 0,$$

which is now written in the form

$$-E_k(\xi_s^k - \dot{x}^k\tau_s) = \frac{d}{dt}\left[\left(L - \dot{x}^k\frac{\partial L}{\partial \dot{x}^k}\right)\tau_s + \frac{\partial L}{\partial \dot{x}^k}\xi_s^k - \Phi_s\right] \quad (s = 1, \ldots, r). \tag{19}$$

This relation represents r identities ($s = 1, \ldots, r$) known as the

Noether Identities *If the fundamental integral of a problem in the calculus of variations is invariant up to an exact differential under a given r-parameter family of transformations (or local Lie group of transformations), then there exist r distinct linear combinations of the Euler–Lagrange expressions which are exact differentials.*

If we set the Euler-Lagrange expressions to zero (necessary conditions for extremal),

$$E_k = \frac{\partial L}{\partial x^k} - \frac{d}{dt}\frac{\partial L}{\partial \dot{x}^k} = 0 \quad (k = 1, \ldots, n), \tag{18'}$$

then the following r expressions hold true:

$$\Omega_s \equiv \left(L - \dot{x}^k \frac{\partial L}{\partial \dot{x}^k}\right)\tau_s + \frac{\partial L}{\partial \dot{x}^k}\xi_s^k - \Phi_s = \text{const} \tag{20}$$

or

$$\Omega_s \equiv -H\tau_s + \frac{\partial L}{\partial \dot{x}^k}\xi_s^k - \Phi_s = \text{const} \quad (s = 1, \ldots, r), \tag{20'}$$

where H is the Hamiltonian function. This conclusion immediately gives rise to the usual formulation of

Noether's Theorem *If the fundamental integral of a problem in the calculus of variations is divergence-invariant under the r-parameter family of transformations* (2), *then r distinct quantities Ω_s defined by* (20) *or* (20′) *are constant along any extremal.*

Since the expressions Ω_s defined by (20) or (20′) are constant whenever $E_k = 0 (k = 1, \ldots, n)$, they are *first integrals* of the differential equations of motion, the Euler–Lagrange equations. *In physical and economic applications, Eqs.* (20) *or* (20′) *are thus interpreted as the "conservation laws" of the system.*

If we define the *canonical momenta* by $p_k \equiv \partial L/\partial \dot{x}^k$, and the *Hamiltonian*, which is in some mechanical systems is the total energy, then (20) or (20′) may be written simply in terms of these canonical variables as

$$-H\tau_s + p_k\xi_s^k - \Phi_s = \quad (s = 1, \ldots, r). \tag{21}$$

Examples

8. The following examples illustrate how the Noether theorem can be applied.

Example 1 (*Conservation of Energy*) Suppose that the potential energy does not explicity depend on the time t: $U = U(x_1, \ldots, x_n)$. Then $L = T - U$, where T is the kinetic energy, is obviously invariant under the time translation $\bar{t} = t + \varepsilon$, $\bar{x}^k = x^k$, which has $\tau_s = 1$, $\xi_s^k = 0$, and $\Phi_s = 0$. Consequently, using (21), a first integral is given by

$$-H = L - p_k\dot{x}^k = -(T + U) = \text{const}.$$

Hence, *if the fundamental integral is invariant under a time translation, then the Hamiltonian, or total energy of the system, is constant—the conservation of energy theorem* (Whittaker [1937]).

Example 2 (*Conservation of Linear Momentum*) Suppose that (1) is absolutely invariant under translation of the form $\bar{t} = t$, $\bar{x}^k = x^k + \varepsilon^k$, which defines an n-parameter group of transformations with generators $\tau_s = 0$, and $\xi_s^k = \delta_s^k$, with $\Phi_s = 0$. This is the case in which the potential energy U is *translation invariant* in the directions of the coordinate axes. The first integrals are given, from (20), by

$$\frac{\partial L}{\partial \dot{x}^k} \delta_s^k = p_k \, \delta_s^k = \text{const} \quad (s = 1, \ \ldots, \ r).$$

That is, the momentum remains unchanged—*conservation of linear momentum.*

Example 3 (*Uniform Motion of the Center of Mass* (Bessel-Hagen, 1921)) Invariance under the velocity transformations, which are sometimes called Galilean transformations, $\bar{t} = t$, $\bar{x}^k + \varepsilon^1 t$, $\bar{y}^k = y^k + \varepsilon^2 t$, $\bar{z}^k = z^k + \varepsilon^3 t$, with the group generators,

$$\begin{aligned}
\tau_s &= \left(\frac{\partial \bar{t}}{\partial \varepsilon^s}\right)_0 = 0, && s = 1, 2, 3, \\
\xi_s^k &= \left(\frac{\partial \bar{x}^k}{\partial \varepsilon^s}\right)_0 = \begin{cases} t, & s = 1, \\ 0, & s = 2, 3, \end{cases} \\
\eta_s^k &= \left(\frac{\partial \bar{y}^k}{\partial \varepsilon^s}\right)_0 = \begin{cases} t, & s = 2, \\ 0, & s = 1, 3, \end{cases} \\
\zeta_s^k &= \left(\frac{\partial \bar{z}^k}{\partial \varepsilon^s}\right)_0 = \begin{cases} t, & s = 3, \\ 0, & s = 1, 2, \end{cases}
\end{aligned}$$

leads to a result which describes how the center of mass of the n particle system moves. These velocity transformations provide us with an example of divergence-invariant, i.e., $\Phi_s \neq 0$. Assume now that the potential energy depends only on the mutual positions of the particles,

$$U = U(\ldots, x^i - x^j, y^i - y^j, z^i - z^j, \ldots).$$

Then it is invariant under the velocity transformations, for we have

$$\begin{aligned}
U &= U(\ldots, \bar{x}^i - \bar{x}^j, \bar{y}^i - \bar{y}^j, \bar{z}^i - \bar{z}^j, \ldots) \\
&= U(\ldots, x^i - x^j, y^i - y^j, z^i - z^j, \ldots).
\end{aligned}$$

Noether's theorem requires that

$$\Omega_s \equiv \sum_k \left(\frac{\partial L}{\partial \dot{x}^k} \xi_s^k + \frac{\partial L}{\partial \dot{y}^k} \eta_s^k + \frac{\partial L}{\partial \dot{z}^k} \zeta_s^k \right) - \Phi_s = \text{const} \quad (s = 1, 2, 3).$$

If the kinetic energy has the form

$$T = \frac{1}{2} \sum_{k=1}^{n} m_k \{ (\dot{x}^k)^2 + (\dot{y}^k)^2 + (\dot{z}^k)^2 \},$$

then

$$\left(\frac{d\bar{x}^k}{d\bar{t}}\right)^2 = \left(\frac{d\bar{x}^k}{dt}\right)^2 = \left(\frac{d}{dt}(x^k + \varepsilon^1 t)\right)^2 = (\dot{x}^k + \varepsilon^1)^2$$
$$= (\dot{x}^k)^2 + 2\varepsilon^1 \dot{x}^k + (\varepsilon^1)^2 = (\dot{x}^k)^2 + \varepsilon^1 \frac{d}{dt}(2x^k) + O(\varepsilon^1).$$

Similar results apply for $(d\bar{y}^k/d\bar{t})^2$ and $(d\bar{z}^k/d\bar{t})^2$. Therefore, we have

$$\bar{T} - T = \frac{1}{2}\sum_k m_k \left\{ \left(\frac{d\bar{x}^k}{d\bar{t}}\right)^2 + \left(\frac{d\bar{y}^k}{d\bar{t}}\right)^2 + \left(\frac{d\bar{z}^k}{d\bar{t}}\right)^2 \right\}$$
$$- \frac{1}{2}\sum_k m_k \left\{ \left(\frac{dx^k}{dt}\right)^2 + \left(\frac{dy^k}{dt}\right)^2 + \left(\frac{dz^k}{dt}\right)^2 \right\}$$
$$= \varepsilon^1 \frac{d}{dt}\sum_k m_k x^k + \varepsilon^2 \frac{d}{dt}\sum_k m_k y^k + \varepsilon^3 \frac{d}{dt}\sum_k m_k z^k.$$

The difference between the Lagrangians in the "barred" and "unbarred" coordinate systems is equal to

$$\bar{L} - L = \varepsilon^1 \frac{d}{dt}\Phi_1 + \varepsilon^2 \frac{d}{dt}\Phi_2 + \varepsilon^3 \frac{d}{dt}\Phi_3,$$

where $\Phi_1 = \sum_k m_k x^k$, $\Phi_2 = \sum_k m_k y^k$, and $\Phi_3 = \sum_k m_k z^k$. Substituting the preceding in the Noether identities, we get

$$t\sum_k m_k \dot{x}^k - \sum_k m_k x^k = \text{const} = -\mathrm{B}_1,$$
$$t\sum_k m_k \dot{y}^k - \sum_k m_k y^k = \text{const} = -\mathrm{B}_2,$$
$$t\sum_k m_k \dot{z}^k - \sum_k m_k z^k = \text{const} = -\mathrm{B}_3.$$

Denoting $M = \sum_k m_k$ as the total mass of the n particles,

$$x_c = \frac{1}{M}\sum_k m_k x^k, \quad y_c = \frac{1}{M}\sum_k m_k y^k, \quad z_c = \frac{1}{M}\sum_k m_k z^k,$$

we observe that (x_c, y_c, z_c) are the coordinates of the center of mass of the system. From Example 2, the potential energy function also allows that the linear momentum is conserved, i.e.,

$$\sum_k m_k \dot{x}^k = \text{const} = A_1, \ \sum_k m_k \dot{y}^k = \text{const} = A_2, \quad \sum_k m_k \dot{z}^k = \text{const} = A_3.$$

Hence, upon combining the above, we get

$$Mx_c = A_1 t + B_1, \quad My_c = A_2 t + B_2, \quad Mz_c = A_3 t + B_3.$$

These show that the center of mass moves uniformly in time or, with constant velocity given by $v_c = (A_1/M, \ A_2/M, \ A_3/M)$. This is *the law of the uniform motion of the center of mass—the center of mass theorem* (Hill [1951]).

Derivation of the Group of Transformation

9. If the Lagrangian and the group transformations are known, then Eq. (17) can be written down directly to obtain the fundamental identities for the system. Now we wish to consider the existence of the group of transformations as an a posteriori notion, and to show how a group can be determined under which a given variational problem is invariant. The fundamental invariance identities can be interpreted as a system of partial differential equations in unknowns τ_s and ξ_s^k for any given Lagrangian function.

Let us assume that a given integral functional (1) is twice continuously differentiable in each of its $2n+1$ arguments, and $x^k(t) \in C^2[t_0, \ t_1]$ for $k = 1, \ \ldots, \ n$. According to the *fundamental invariance identities* (17), there exist the r relations (17), where τ_s and ξ_s^k are the generators of the group

$$\bar{t} = t + \tau_s(t,x)\varepsilon^s, \quad \bar{x}^k = x^k + \xi_s^k(t,x)\varepsilon^s \quad (s = 1, \ \ldots, \ r). \tag{5'}$$

Upon expanding the total derivatives in (17), we get

$$\begin{aligned}&\frac{\partial L}{\partial t}\tau_s + \frac{\partial L}{\partial x^k}\xi_s^k + \frac{\partial L}{\partial \dot{x}^k}\left(\frac{\partial \xi_s^k}{\partial t} + \frac{\partial \xi_s^k}{\partial x^j}\dot{x}^j - \dot{x}^k\left(\frac{\partial \tau_s}{\partial t} + \dot{x}^j\frac{\partial \tau_s}{\partial x^j}\right)\right)\\ &\quad + L\left(\frac{\partial \tau_s}{\partial t} + \frac{\partial \tau_s}{\partial x^j}\dot{x}^j\right) = \frac{d\boldsymbol{\Phi}_s}{dt} \quad (s = 1, \ \ldots, \ r).\end{aligned} \tag{22}$$

Altering our viewpoint, Eq. (22) are identities in $(t, \ x^k)$ for arbitrary directional arguments $\dot{x}^k$. That is to say, τ_s and ξ_s^k depend only on t and x^k but not on $\dot{x}^k$. Hence we can regard (22) as a set of partial differential equations in unknowns τ_s and ξ_s^k. *Due to the arbitrariness of the* $\dot{x}^k$, (22) *will be further reduced to a system of first-order partial differential equations in* τ_s *and* ξ_s^k *by equating to zero the coefficients of the powers of the* $\dot{x}^k$. The general method of the derivation of a group is then: *If this reduced system admits a solution, we have determined a group under which the fundamental integral is invariant, from which conservation laws will be derived by the Noether theorem.*

It may be best to illustrate this general method using the following Lagrangian which leads to the so-called Emden equation (see H. T. Davis [1962, p. 371 ff])

$$L = t^2(\frac{1}{2}\dot{x}^2 - \frac{1}{6}x^6).$$

The Emden equation is the Euler–Lagrange equation of L, i.e.,

$$\frac{\partial L}{\partial x} - \frac{d}{dt}\left(\frac{\partial L}{\partial \dot{x}}\right) = 0 \quad \text{or} \quad t\ddot{x} + 2\dot{x} + tx^5 = 0.$$

We look for a one-parameter group of transformations $\bar{t} = t + \tau(t,x)\varepsilon$, $\bar{x} = x + \xi(t, \ x)\varepsilon$ under which $J(x) = \int_{t_0}^{t_1} L\,dt$ is invariant. Calculating $\partial L/\partial t = t(\dot{x}^2 - \frac{1}{3}x^6)$, $\partial L/\partial x = -t^2x^5$, and $\partial L/\partial \dot{x} = t^2\dot{x}$ and substituting these quantities into the invariance identity (22) (with $s = 1, k = 1$, and $d\boldsymbol{\Phi}/dt = 0$), we obtain

$$t\left(\dot{x}^2 - \frac{x^6}{3}\right)\tau + (-t^2x^5)\xi + t^2\dot{x}\left(\frac{\partial\xi}{\partial t} + \frac{\partial\xi}{\partial x}\dot{x} - \dot{x}\frac{\partial\tau}{\partial t} - \dot{x}^2\frac{\partial\tau}{\partial x}\right)$$
$$+ t^2\left(\frac{\dot{x}^2}{2} - \frac{x^6}{6}\right)\left(\frac{\partial\tau}{\partial t} + \dot{x}\frac{\partial\tau}{\partial x}\right) = 0.$$

Or by dividing with t and collecting the coefficients of the powers of $\dot{x}$, namely, $\dot{x}^0$, $\dot{x}^1$, $\dot{x}^2$, and $\dot{x}^3$, we get

$$\left(-\frac{x^6}{3}\tau - tx^5\xi - \frac{tx^6}{6}\frac{\partial\tau}{\partial t}\right)\dot{x}^0 + \left(t\frac{\partial\xi}{\partial t} - \frac{tx^6}{6}\frac{\partial\tau}{\partial x}\right)\dot{x}^1$$
$$+\left(\tau + t\frac{\partial\xi}{\partial x} - \frac{t}{2}\frac{\partial\tau}{\partial t}\right)\dot{x}^2 + \left(-\frac{t}{2}\frac{\partial\tau}{\partial x}\right)\dot{x}^3 = 0.$$

The coefficients of the powers of $\dot{x}$ must be equal to zero due to the arbitrariness of directional argument $\dot{x}$ and due to the independence of τ and ξ from $\dot{x}$. Hence the following system of partial differential equations for τ and ξ must hold:

$$\text{(i)}\ \frac{x}{3}\tau + t\xi + \frac{tx}{6}\frac{\partial\tau}{\partial t} = 0, \quad \text{(ii)}\ \frac{\partial\xi}{\partial t} - \frac{x^6}{6}\frac{\partial\tau}{\partial x} = 0,$$
$$\text{(iii)}\ \tau + \frac{\partial\xi}{\partial x}t - \frac{t}{2}\frac{\partial\tau}{\partial t} = 0, \quad \text{(iv)}\ \frac{\partial\tau}{\partial x}\frac{t}{2} = 0.$$

Since t is not equal to zero, (iv) immediately gives that τ is a function of only t, $\tau = \tau(t)$, $\partial\tau/\partial x = 0$. As a result, from (ii) we have $\partial\xi/\partial t = 0$ or $\xi = \xi(x)$. Using these, the foregoing conditions reduce to

$$\frac{x}{3}\tau + t\xi + \frac{tx}{6}\frac{d\tau}{dt} = 0, \quad \tau + t\frac{d\xi}{dx} - \frac{t}{2}\frac{d\tau}{dt} = 0,$$

which are simple ordinary differential equations. By differentiating the first equation with respect to x and noting that τ is independent of x, we get

$$\frac{\tau}{3} + t\frac{d\xi}{dx} + \frac{t}{6}\frac{d\tau}{dt} = 0, \quad \tau + t\frac{d\xi}{dx} - \frac{t}{2}\frac{d\tau}{dt} = 0.$$

Elimination of $t\ d\xi/dx$ from these two equations gives

$$-\frac{2}{3}\tau + \frac{2}{3}t\frac{d\tau}{dt} = 0.$$

Then, τ is immediately derived as $\tau = ct$ or by setting $c = 1$, $\tau = t$. Substituting this in the first equation, we get $\xi = -\frac{1}{2}x$. From Noether's theorem, we obtain the first integral or constant of the motion,

$$\boldsymbol{\Omega} \equiv \left(L - \dot{x}\frac{\partial L}{\partial \dot{x}}\right)\tau + \frac{\partial L}{\partial \dot{x}}\xi \equiv \frac{t^3x^6}{6} + \frac{t^3\dot{x}^2}{2} + \frac{t^2x\dot{x}}{3} = \text{const} = C.$$

This is the *conservation law* for the systems associated with the Emden equation.

3. Conservation Laws in Simple Models of the Ramsey Type[8]

Simple Models of the Ramsey Type

10. We are now in a position to derive conservation laws in dynamic economic systems. We begin with simple growth models of the Ramsey type. These models are characterized with the *linear* transformation function (linear production possibility frontier) between consumption and capital accumulation, i.e.,

$$c = c(\dot{k},\ k) = f(k) - nk - \dot{k}, \quad n \geq 0, \tag{23}$$

where c is the consumption (per capita C/L), $f(k) = Y/L$ the neoclassical (aggregate) production function (per capita) with $f'(k) > 0,\ f''(k) < 0,\ k = K/L$ is the capital–labor ratio, n the sum of the constant population growth rate and the depreciation rate, and $\dot{k} = dk/dt$ the rate of capital (per capita) accumulation. Although Ramsey's original work [1928] considers the case of $n = 0$ and $k = K,\ c = C, \dot{k} = \dot{K}$, and $f = Y$, the qualitative properties of his model are basically the same as those developed later by Samuelson and Solow [1956], Cass [1965], Koopmans [1965], Burmeister and Dobell [1970], Samuelson [1965], Sato and Davis [1971], Shell [1967], and others. The basic feature of these models is that the production possibility frontier, as depicted by Fig. 3, is linear, i.e.,

$$\partial^2 c/\partial \dot{k}^2 \equiv 0. \tag{24}$$

The objective of society is to maximize the social welfare functional, i.e.,

$$J(k) = \int_0^T e^{-\rho t} U(c) dt \rightarrow \max \tag{25}$$

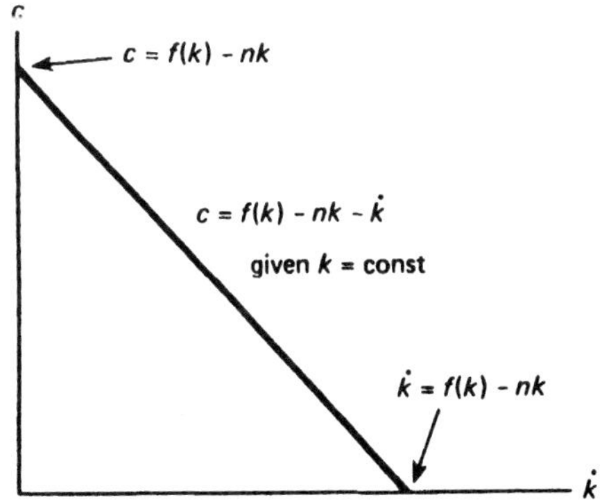

FIGURE 3. Linear Production Frontier

[8] The author gratefully acknowledges helpful assistance and suggestions rendered by Professor Takayuki Nôno.

under condition (23), together with appropriate initial conditions. In (25), U is a utility (or welfare) function with the usual properties $U' > 0$, $U'' < 0$, and $\rho \geqq 0$ is the fixed discount rate. The terminal time T may or may not be finite. In many cases we may assume that it is infinite, $T = +\infty$. Substituting (23) into (25), we have the standard calculus of variation (or optimal control) problem (25), where c is defined by (23).

Here we have the Lagrangian

$$L(t,\ k(t),\ \dot{k}(t)) = e^{-\rho t} U[g(k(t)) - \dot{k}(t)], \tag{26}$$

where

$$g(k) = f(k) - nk. \tag{27}$$

We look for a family of transformations,[9]

$$\bar{t} = t + \varepsilon\tau(t,\ k), \tag{28-i}$$

$$\varepsilon = \text{parameter},;$$

$$\bar{k} = k + \varepsilon\xi(t,\ k), \tag{28-ii}$$

under which (25) is *dynamically invariant*. In the invariance identity [Eq. (17) or (22)], we first assume $d\Phi/dt = 0$. Calculating $\partial L/\partial t = -\rho e^{-\rho t}U$, $\partial L/\partial k = e^{-\rho t}U'g'$, and $\partial L/\partial\dot{k} = -e^{-\rho t}U'$, and substituting these in the *invariance identity*, we get

$$U\left(\frac{d\tau}{dt} - \rho\tau\right) + U'\left(g'\xi - \frac{d\xi}{dt} + \dot{k}\frac{d\tau}{dt}\right) = 0. \tag{29}$$

This is the *fundamental invariance equation for the simple Ramsey models*. In order that (25) may be invariant under *any arbitrary form* of U and that conservation laws may hold for *any* utility function, (29) must imply

$$\frac{d\tau}{dt} - \rho\tau = 0, \tag{29-i$'$}$$

$$g'\xi - \frac{d\xi}{dt} + \dot{k}\frac{d\tau}{dt} = 0. \tag{29-ii$'$}$$

On using $d\tau/dt = \partial\tau/\partial t + \dot{k}\partial\tau/\partial k$ and $d\xi/dt = \partial\xi/\partial t + \dot{k}\partial\xi/\partial k$, (29$'$) is equal to

$$\frac{\partial\tau}{\partial t} + \dot{k}\frac{\partial\tau}{\partial k} - \rho\tau = 0, \tag{29-i$''$}$$

$$g'\xi - \left(\frac{\partial\xi}{\partial t} + \dot{k}\frac{\partial\xi}{\partial k}\right) + \dot{k}\left(\frac{\partial\tau}{\partial t} + \dot{k}\frac{\partial\tau}{\partial k}\right) = 0. \tag{29-ii$''$}$$

[9] We may give economic interpretations to (28). The time transformation may be considered as the transformation of "objective time" to "subjective time" (Samuelson [1976]), where the subjective time depends not only on the objective time t, but also on the wealth k. The second equation of (28) is nothing but the technical change transformation depending also on the objective time and the wealth effect.

Collecting the coefficients of the powers of $\dot{k}$, we have

$$\left(\frac{\partial\tau}{\partial t} - \rho\tau\right)\dot{k}^0 + \frac{\partial\tau}{\partial k}\dot{k}^1 = 0,$$

$$\left(g'\xi - \frac{\partial\xi}{\partial t}\right)\dot{k}^0 + \left(-\frac{\partial\xi}{\partial k} + \frac{\partial\tau}{\partial t}\right)\dot{k}^1 + \frac{\partial\tau}{\partial k}\dot{k}^2 = 0.$$

On setting these coefficients equal to zero, the invariance condition becomes

$$\frac{\partial\tau}{\partial t} - \rho\tau = 0, \tag{29-i'''}$$

$$\frac{\partial\tau}{\partial k} = 0, \tag{29-ii'''}$$

$$g'\xi - \frac{\partial\xi}{\partial t} = 0, \tag{29-iii'''}$$

$$-\frac{\partial\xi}{\partial k} + \frac{\partial\tau}{\partial t} = 0. \tag{29-iv'''}$$

From (29-ii'''), we immediately conclude that $\tau = \tau(t)$, and from (29-i''')

$$\tau = ae^{\rho t} \quad (a =). \tag{30-i}$$

From (29-iv''') and using (30-i), $\partial\xi/\partial k = \partial\tau/\partial t = a\rho e^{\rho t}$. Integrating **(29-iii''')** partially with respect to t, we derive

$$\xi = A(k)e^{g'(k)t}. \tag{30-ii}$$

But from (29-iv'''), $\partial\xi/\partial k = \partial\tau/\partial t$, and from (30-i), $\tau = ae^{\rho t}$, we can show that $\partial^2\xi/\partial k^2 \equiv 0$. By differentiating ξ with respect to k once,

$$\frac{\partial\xi}{\partial k} = e^{g'(k)t}[A(k)g''(k)t + A'(k)],$$

and twice,

$$\frac{\partial^2\xi}{\partial k^2} = g''(k)te^{g'(k)t}[A(k)g''(k)t + A'(k)] + e^{g'(k)t}[(A(k)g''(k))'t + A''(k)],$$

and using the fact $\partial^2\xi/\partial k^2 \equiv 0$, we get

$$t^2g''(k)^2A(k) + t[g''(k)A'(k) + (g''(k)A(k))'] + A''(k) = 0.$$

The foregoing requires that

$$A''(k) = 0, \tag{31-i}$$

$$g''(k)^2A(k) = 0, \tag{31-ii}$$

$$g''(k)A'(k) + (g''(k)A(k))' = 0. \tag{31-iii}$$

There are two cases depending upon $g''(k) \neq 0$ or $g''(k) = 0$.

Case 1 ($g''(k) \neq 0 \Leftrightarrow f''(k) \neq 0$ from (27), i.e., $f(k) \neq \alpha k + \beta$) Equation (31-ii) requires that $A(k) = 0$, which implies from (30-ii), $\xi = 0$ (no conservation law with respect to $\bar{k}$).

Also, from (29-iv‴) we must have $\partial\tau/\partial t = 0$. Substituting this into

$$\partial\tau/\partial t - \rho\tau = 0, \tag{29-i'''}$$

we have $\rho\tau = 0$, which means

$$\rho \neq 0, \quad \tau = 0 \quad \text{(no conservation law with respect to } \bar{t}\text{)}, \tag{32-i}$$

$$\rho = 0, \quad \tau = \quad \text{(time translation)}. \tag{32-ii}$$

Case 2 ($g''(k) = 0 \Leftrightarrow f''(k) = 0$, i.e., $f(k) = \alpha k + \beta$, $\alpha > 0$ or $g(k) = (\alpha - n)$ $k + \beta$) Here we have from (31-i), $A''(k) = 0$, or $A(k) = bk + d$. Substituting this into (30-ii), we get $\xi = (bk + d)e^{(\alpha - n)t}$, $g' = (\alpha - n)$. We must now determine the consistent values of α and b that satisfy (29‴). It is easily seen that (29-iii‴) is always met. Hence (29-iv‴) is the only restriction that must be satisfied, which relates ξ with τ. Thus

$$\partial\xi/\partial k = be^{(\alpha - n)t} = \rho a e^{\rho t} = \partial\tau/\partial t.$$

Therefore, we must have $b = \rho a$ and $\rho = \alpha - n$ or $\alpha = \rho + n$, which gives the final form of the generator of the group as

$$\tau = ae^{\rho t}, \tag{30-i'}$$

$$\xi = (a\rho k + d)e^{\rho t} \tag{30-ii'}$$

for $g(k) = \rho k + \beta = f(k) - nk = \alpha k + \beta - nk$, or $f(k) = g(k) + nk = \rho k +$ $\beta + nk = (\rho + n)k + \beta$($a$, β, d are arbitrary constants).

Case 1 (General Production Function $f(k) \neq \alpha k + \beta$) (i) $\rho \neq 0$ (positive discount rate $\rho > 0$)

$$\xi = 0, \quad \tau = 0, \quad \text{no conservation laws.}$$

(ii) $\rho = 0$ (zero discount rate)

$$\xi = 0, \quad \tau = \text{const} = 1, \quad \text{time translation,}$$

$$\boldsymbol{\Omega} = L\tau + \frac{\partial L}{\partial \dot{k}}(\xi - \dot{k}\tau) = L\tau - \frac{\partial L}{\partial \dot{k}}\dot{k}\tau,$$

or

$$\boldsymbol{\Omega} = U + U'\dot{k} = \text{Hamiltonian} = \text{const} \quad (\tau = 1). \tag{33}$$

Case 2 (Linear Technology $f(k) = (\rho + n)k + \beta$) (i) $\tau = ae^{\rho t}$,

(ii) $\xi = (a\rho k + d)e^{\rho t}$ for

$$L = e^{-\rho t} U(\rho k + \beta - \dot{k});$$

$$\begin{aligned}
\boldsymbol{\Omega} &= L\tau + \frac{\partial L}{\partial \dot{k}}(\xi - \dot{k}\tau) = \left(L - \frac{\partial L}{\partial \dot{k}}\dot{k}\right)\tau + \frac{\partial L}{\partial \dot{k}}\xi = -\text{Hamiltonian} \times \tau \\
&\quad + \frac{\partial L}{\partial \dot{k}}\xi = \text{const} = e^{-\rho t} U a e^{\rho t} - e^{-\rho t} U'[(a\rho k + d)e^{\rho t} - \dot{k} a e^{\rho t}] \\
&= aU - U'[a(g(k) - \dot{k}) + d - a\beta] \quad (g(k) = \rho k + \beta),
\end{aligned}$$

or

$$\boldsymbol{\Omega} = a[U(c) - U'(c)c] + (a\beta - d)U'(c) = \text{const} \quad (g(k) - \dot{k} = c). \tag{34}$$

Up to this point we have sought conservation laws under the definition of absolute invariance (Definition 1). We now look for laws under the more general definition of divergence-invariant (Definition 2). The invariance equation (17) requires that

$$\frac{\partial L}{\partial t}\tau + \frac{\partial L}{\partial k}\xi + \frac{\partial L}{\partial \dot{k}}\left(\frac{d\xi}{dt} - \dot{k}\frac{d\tau}{dt}\right) + L\frac{d\tau}{dt} = \frac{d\boldsymbol{\Phi}}{dt},$$

where $\boldsymbol{\Phi} = \boldsymbol{\Phi}(t, k)$ and $L = e^{-\rho t} U(g(k) - \dot{k})$. Substituting $\partial L/\partial t = -\rho e^{-\rho t} U$, $\partial L/\partial k = e^{-\rho t} U' g'$, $\partial L/\partial \dot{k} = -e^{-\rho t} U'$ into the foregoing and writing

$$\frac{d\xi}{dt} = \frac{\partial \xi}{\partial t} + \dot{k}\frac{\partial \xi}{\partial k}, \quad \frac{d\tau}{dt} = \frac{\partial \tau}{\partial t} + \dot{k}\frac{\partial \tau}{\partial k}, \quad \text{and} \quad \frac{d\boldsymbol{\Phi}}{dt} = \frac{\partial \boldsymbol{\Phi}}{\partial t} + \dot{k}\frac{\partial \boldsymbol{\Phi}}{\partial k},$$

we get

$$\begin{aligned}
&e^{-\rho t}\left[U\left(\frac{\partial \tau}{\partial t} - \rho\tau + \frac{\partial \tau}{\partial k}\dot{k}\right) + U'\left\{g'\xi - \frac{\partial \xi}{\partial t} - \left(\frac{\partial \xi}{\partial k} - \frac{\partial \tau}{\partial t}\right)\dot{k} + \frac{\partial \tau}{\partial k}\dot{k}^2\right\}\right] \\
&\quad = \frac{\partial \boldsymbol{\Phi}}{\partial t} + \dot{k}\frac{\partial \boldsymbol{\Phi}}{\partial k}.
\end{aligned}$$

Differentiating the above partially twice with respect to $\dot{k}$, we obtain

$$\begin{aligned}
&U'''\left\{\left(g'\xi - \frac{\partial \xi}{\partial t}\right) - \left(\frac{\partial \xi}{\partial k} - \frac{\partial \tau}{\partial t}\right)\dot{k} + \frac{\partial \tau}{\partial k}\dot{k}^2\right\} \\
&\quad + U''\left\{\left(\frac{\partial \tau}{\partial t} - \rho\tau\right) + 2\left(\frac{\partial \xi}{\partial k} - \frac{\partial \tau}{\partial t}\right) - 3\frac{\partial \tau}{\partial k}\dot{k}\right\} = 0.
\end{aligned}$$

In order that the conservation laws may hold for any arbitrary utility function, we must have, after setting the coefficients of the powers of $\dot{k}$ to zero,

$$\begin{aligned}
&g'\xi - \frac{\partial \xi}{\partial t} = 0, \quad \frac{\partial \tau}{\partial k} = 0 \\
&\frac{\partial \xi}{\partial k} - \frac{\partial \tau}{\partial t} = 0, \quad \frac{\partial \tau}{\partial t} - \rho\tau + 2\left(\frac{\partial \xi}{\partial k} - \frac{\partial \tau}{\partial t}\right) = 0.
\end{aligned}$$

This implies that

$$\frac{\partial \tau}{\partial t} - \rho\tau = 0 \quad \left(\text{from} \frac{\partial \xi}{\partial k} - \frac{\partial \tau}{\partial t} = 0\right), \tag{35-i}$$

$$\frac{\partial \tau}{\partial k} = 0, \tag{35-ii}$$

$$g'\xi - \frac{\partial \xi}{\partial t} = 0, \tag{35-iii}$$

$$\frac{\partial \xi}{\partial k} - \frac{\partial \tau}{\partial t} = 0. \tag{35-iv}$$

But *this is exactly identical with* (29‴), *the case of absolute invariance. Hence there exist no additional conservation laws, even if we allow for a more general definition of invariance.*

Economic Interpretation

11. We first give an economic interpretation to Noether's theorem [Eq. (20) or (21)] when it is applied to a typical welfare maximization problem with n state variables.

If a welfare function is dynamically invariant under an r-parameter family of transformations resulting from technical change and/or taste change, then the following quantities are constant along any optimal path for the entire planning period:

$$(\text{Hamiltonian})\begin{pmatrix}\text{infinitesimal}\\ \text{transformations}\\ \text{of time}\end{pmatrix} + \sum_{i=1}^{n}\begin{pmatrix}\text{implicit}\\ \text{price of}\\ x_i\end{pmatrix}\begin{pmatrix}\text{infinitesimal}\\ \text{transformations}\\ \text{of } x_i \text{ due to}\\ \text{technical}\\ \text{(taste) change}\end{pmatrix} + (\text{null term})$$
$$= \text{const}$$
$$= \begin{pmatrix}\text{measure of}\\ \text{welfare per}\\ \text{infinitesimal}\\ \text{change of time}\end{pmatrix} + \sum_{i=1}^{n}\begin{pmatrix}\text{value (effect)}\\ \text{of technical}\\ \text{(taste) change}\end{pmatrix} + (\text{null term})$$
$$= \text{const.}$$

Turning our attention to the specific conservation laws derived in the previous section, we observe that conservation laws for the simple Ramsey models crucially depend on whether the discount rate is zero or not, and/or whether technology is linear or not. When production technology is of the general neoclassical type, $f(k) > 0, f'(k) > 0, f''(k) < 0$, $k > 0$, and when the discount rate is positive, then there exists *no* conservation law for an arbitrary form of utility function. On the other hand, when technology is linear with the marginal product equal to the exogenously determined rate of $\rho + n$, there exists an *invariance identity* $U(c) = \text{const}$, i.e., $c = \text{const}$ for every period.

When the discount rate is zero, that is to say, the Lagrangian does not contain t explicitly, the value of the Hamiltonian is, as is expected, constant. What is important here, however, is that *the constancy of the Hamiltonian is the only conservation law operating in the system when* $\rho = 0$. The economic interpretation of the constancy of the Hamiltonian [Eq. (33)] or the invariance under time translation (32-ii) is that the sum of the value of consumption ($U(c)$) and the value of investment ($q\dot{k}$, $q = U'$), which is the measure of welfare in terms of consumption, remains constant. That is to say, the value of income measured in terms of consumption ($U' = q$) does not change over the planning period.

The second case of singularity requires a further explanation. First, the conservation law (34) does *not* hold for *all* linear technology. It holds only for a special type of linear technology with the marginal product of capital identically equal to $\rho + n$, the sum of the discount rate and the population growth (and the Harrod neutral technical progress) rate. Since there are *two* parameters a and d in the infinitesimal transformation (30-i′) and (30-ii″), there are *two* quantities that are constant (Noether's theorem). Thus we have

$$\tau_1 = ae^{\rho t}, \quad d = 0, \quad \text{and} \quad \tau_2 = 0, \quad a = 0,$$
$$\xi_1 = a\rho k e^{\rho t}, \ d = 0, \qquad \xi_2 = de^{\rho t}, \ a = 0.$$

Corresponding to these cases using (34) we obtain

$$\Omega_1 = a[U(c) - U'(c)c + \beta U'(c)] = \text{const} \tag{34-i}$$

and

$$\Omega_2 = -dU'(c) = \text{const}. \tag{34-ii}$$

Looking at Ω_2, we immediately conclude that $U'(c) = \text{const}$ or $c = \text{const}$, that is to say, *the optimally controlled per capita consumption must remain unchanged during the planning horizon.* The conditions on Ω_1 do not say more than this, for if $c = \text{const}$, naturally, Ω_1, which is a function of only c, must also be constant. Hence, as the economic conservation law, we derive the constancy of *per capita consumption.*

This singular case of linear technology corresponds to the Arrow–Kurz example (Kurz [1968a]), when the marginal product in the Arrow–Kurz example $\alpha = \rho$ in our case. (Here our case applies to any utility function.) The optimal control program is to maximize

$$\int_0^\infty e^{-\rho t} U(c) dt$$

subject to $k = (\rho + n)k - nk + \beta - c = \rho k + \beta - c$, $U' > 0$, $U'' < 0$, $k > 0$.

The Euler–Lagrange equation (or the maximum principle) requires that

$$k = \rho k + \beta - c(q) \quad \text{and} \quad \dot{q} = (\rho - \rho)q = 0, \quad U'(c) = q,$$

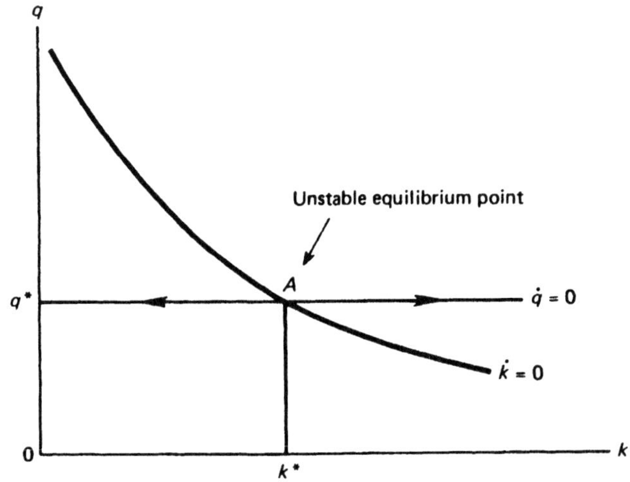

FIGURE 4. Phase Diagram for Unstable Equilibrium

where $q = U'(c)$ is the current value of the implicit price of investment. The phase diagram is shown in Fig. 4. Given the planning horizon, $U'(c) = q = \text{const} > 0$, because of the conservation law, $q = q^*$, while k must always be increasing determined by $\dot{k} = \rho k + \beta - c^*(q^*)$ or $k(t) = [k(0) - k^*]e^{\rho t} + k^*$, where $k^* = (c^* - \beta)/\rho$ and $k(0)$ is the initial condition. That is to say, k must be eventually increasing or decreasing at the rate of the discount rate ρ. Here, as Kurz demonstrates, the optimal program has the totally unstable feature.

The following theorem summarizes the results up to this point.

Theorem 1 (*Conservation Laws for Simple Models of the Ramsey Type*)

(i) *When the discount rate ρ is zero, there exists only one conservation law that the value of welfare measured in terms of consumption and investment remains constant, i.e.,*

$$H = U(c) + q\dot{k} = \text{const}, \quad q = U'.$$

This is an alternative statement of the Ramsey rule in Section 2.

(ii) *When the discount rate ρ is positive, there exists* no *conservation law unless the production technology is a special linear type $f(k) = \alpha k + \beta = (\rho + n)k + \beta$, $\alpha = \rho + n$.*

(iii) *When the production function is of the form $f(k) = \alpha k + \beta = (\rho + n)k + \beta$, $\alpha = \rho + n$, $\rho > 0$, the value of per capita consumption c is always constant. Hence the (current value) implicit price of investment $q = U'$, and therefore, the utility level itself remains unchanged. The discounted welfare is decreasing at the constant rate ρ.*

(iv) *These results hold true even when the more general definition of invariance with nullity is adopted.*

INDEX

Page numbers followed by f indicate figures.

Printed in the United Kingdom
by Lightning Source UK Ltd.
115325UKS00008B/238

9 780387 260556